Conceptual Geometry
of
Straight Line

A Companion to S. L. Loney's Co-ordinate Geometry

By

Chandra Shekhar Kumar
Integrated M. Sc. in Physics, IIT Kanpur, India
CEO & Co-Founder, Ancient Kriya Yoga Mission
CTO & Co-Founder, Ancient Science Publishers

Ancient Science Publishers

TeX is a trademark of the American Mathematical Society.
METAFONT is a trademark of Addison-Wesley.

Care has been taken in the preparation of this book, but makes no expressed or implied warranty of any kind and assumes no responsibility for errors or omissions. No liability is assumed for incidental or consequential damages in connection with or arising out of the use of the information contained herein.

For comments, suggestions and or feedback, send mail to :
ancientsciencepublishers@gmail.com

Copyright ©2018 Chandra Shekhar Kumar

All rights reserved. This work is protected by copyright and permission must be obtained prior to any prohibited reproduction, storage in a retrieval system, or transmission in any form or by any means, electronic, mechanical, photocopying, recording, or likewise unless stated otherwise.

ISBN-13: 978-1986834322
ISBN-10: 1986834328

Ancient Science Publishers, # 4, Suresh Prasad Singh, Sardar Patel Path, Boring Road, Patna, Bihar 800013, India.

Highly Recommended Books for Self Study & Competitions

Author : Chandra Shekhar Kumar

Conceptual Kinematics
A Companion to I. E. Irodov's Problems in General Physics

Conceptual Geometry
A Companion to S. L. Loney's Co-ordinate Geometry

Conceptual Trigonometry Part I
A Companion to S. L. Loney's Plane Trigonometry Part I

Conceptual Trigonometry Part II
A Companion to S. L. Loney's Plane Trigonometry Part II

Conceptual Dynamics
A Companion to S. L. Loney's Elements of Dynamics

Conceptual Statics
A Companion to S. L. Loney's Elements of Statics

Conceptual Particle Dynamics
A Companion to S. L. Loney's Dynamics of A Particle

Conceptual Rigid Body Dynamics
A Companion to S. L. Loney's Dynamics of Rigid Bodies

Conceptual School Geometry
A Companion to Hall & Stevens' School Geometry

Conceptual School Algebra
A Companion to Hall & Knight's Elementary Algebra

Problems and Solutions in Plane Trigonometry
by
Isaac Todhunter & Neeru Singh

Solutions of the Examples in Higher Algebra
by
H. S. Hall, S. R. Knight, Neeru Singh
& C. S. Kumar

Questions and Problems in School Physics
A Companion to I. E. Irodov's Problems in General Physics
by
Lev Tarasov, Aldina Tarasova
& Chandra Shekhar Kumar

Calculus
Basic Concepts for High Schools
by
Lev Tarasov & Chandra Shekhar Kumar

General Methods for Solving Physics Problems
A Companion to I. E. Irodov's Problems in General Physics
by
B. S. Belikov & Chandra Shekhar Kumar

Preface

Back in 1990, solving the problems and exercises given in the text-book of **Co-ordinate Geometry** by **S. L. Loney** had a terrorizing effect on me, irrespective of the outcome of the countless hours, full of perspiration and inspiration, laced with joy and surrendering to the sheer beauty and elegance of each problem, sub-problem, ... woven with multi-concepts.. Whenever stuck, I used to revise the concepts embedded in the text-book and related references, monographs, take a break and start all over... an irresistible journey... back n forth between the classics of Loney and others.

As time grew, I ended up stocking a huge pile of sheets comprising of my notes as an endeavor to solve and devour the entire book and beyond (needless to mention that I laid my hands on everything I could in my pursuit).

Somewhere in 2005, I started collating and organizing my notes to instill coherence and capture the elegance in the flow.

The present work is an outcome of this pursuit, which will serve as a complete guide to private students reading the subject with few or no opportunities of instruction. This will save the time and lighten the work of Teachers as well. This book helps in acquiring a better understanding of the basic principles of Straight Line and in revising a large amount of the subject matter quickly. Care has been taken, as in the forthcoming ones, to present the solutions with multi-concepts and beyond in a simple natural manner, in order to meet the difficulties which are most likely to arise, and to render the work intelligible and instructive.

This work contains several variations of problems, solutions, methods, approaches to enrich, strengthen and enliven the inherent multi-concepts.

Ancient Science Publishers *Chandra Shekhar Kumar*
March, 2018.

List of Chapters

Preface		i
1	**Introduction**	**1**
	1.1 Algebraic Results	1
2	**Coordinates : Lengths of Straight Lines and Areas of Triangles**	**3**
	2.1 Coordinates	3
	2.2 Length of Straight Line	4
	2.3 Areas of Triangles	18
	2.4 Polar Coordinates	23
3	**Locus : Equation to a Locus**	**35**
4	**The Straight Line : Rectangular Coordinates**	**43**
	4.1 Equations, Slope and Intercepts	43
	4.2 Angles between straight lines	56
	4.3 Lengths of Perpendiculars	66
	4.4 Bisectors of angles between straight lines	69
5	**The Straight Line : Polar Equations, Oblique Coordinates and Loci**	**99**
	5.1 Oblique Coordinates	99
	5.2 Straight lines passing through fixed pts.	107
	5.3 Loci	130
6	**On Equations Representing Two Or More Straight Lines**	**145**
	6.1 Multiple Straight Lines and Included Angles	145
	6.2 General Equation of The Second Degree	152
	6.3 Equations Representing Isolated Points	162

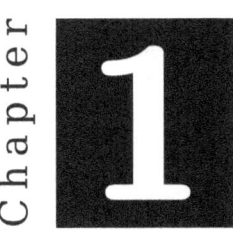

Chapter 1

Introduction

1.1 Algebraic Results

Prove that

§ Problem 1.1.1. $\begin{vmatrix} 2 & -3 \\ 4 & 8 \end{vmatrix} = 28.$ ◊

§§ Solution. $\begin{vmatrix} 2 & -3 \\ 4 & 8 \end{vmatrix} = 2 \times 8 - (-3) \times 4 = 16 + 12 = 28.$ ∎

§ Problem 1.1.2. $\begin{vmatrix} -6 & 7 \\ -4 & -9 \end{vmatrix} = 82.$ ◊

§§ Solution. $\begin{vmatrix} -6 & 7 \\ -4 & -9 \end{vmatrix} = (-6) \times (-9) - 7 \times (-4) = 54 + 28 = 82.$ ∎

§ Problem 1.1.3. $\begin{vmatrix} 5 & -3 & 7 \\ -2 & 4 & -8 \\ 9 & 3 & -10 \end{vmatrix} = -98.$ ◊

§§ Solution. $\begin{vmatrix} 5 & -3 & 7 \\ -2 & 4 & -8 \\ 9 & 3 & -10 \end{vmatrix} = 5 \times \begin{vmatrix} 4 & -8 \\ 3 & -10 \end{vmatrix} - (-3) \times \begin{vmatrix} -2 & -8 \\ 9 & -10 \end{vmatrix} + 7 \times \begin{vmatrix} -2 & 4 \\ 9 & 3 \end{vmatrix}$

$= 5 \times \{-40 + 24\} + 3 \times \{20 + 72\} + 7 \times \{-6 - 36\}$
$= 5 \times (-16) + 3 \times 92 + 7 \times (-42)$
$= -80 + 276 - 294 = -98.$ ∎

§ Problem 1.1.4. $\begin{vmatrix} 9 & 8 & 7 \\ 6 & 5 & 4 \\ 3 & 2 & 1 \end{vmatrix} = 0$ ◊

1.1. Algebraic Results

§§ Solution. $\begin{vmatrix} 9 & 8 & 7 \\ 6 & 5 & 4 \\ 3 & 2 & 1 \end{vmatrix} = 9 \times \begin{vmatrix} 5 & 4 \\ 2 & 1 \end{vmatrix} - 8 \times \begin{vmatrix} 6 & 4 \\ 3 & 1 \end{vmatrix} + 7 \times \begin{vmatrix} 6 & 5 \\ 3 & 2 \end{vmatrix}$

$= 9 \times \{5 \times 1 - 4 \times 2\} - 8 \times \{6 \times 1 - 4 \times 3\} + 7 \times \{6 \times 2 - 5 \times 3\}$
$= 9 \times \{5 - 8\} - 8\{6 - 12\} + 7\{12 - 15\}$
$= 9 \times (-3) - 8 \times (-6) + 7 \times (-3) = -27 + 48 - 21 = 0.$ ∎

§ Problem 1.1.5. $\begin{vmatrix} -a & b & c \\ a & -b & c \\ a & b & -c \end{vmatrix} = 4abc.$ ◊

§§ Solution. $\begin{vmatrix} -a & b & c \\ a & -b & c \\ a & b & -c \end{vmatrix} = -a \times \begin{vmatrix} -b & c \\ b & -c \end{vmatrix} - b \times \begin{vmatrix} a & c \\ a & -c \end{vmatrix} + c \times \begin{vmatrix} a & -b \\ a & b \end{vmatrix}$

$= -a \times \{(-b) \times (-c) - c \times b\} - b \times \{a \times (-c) - c \times a\} + c \times \{a \times b - (-b) \times a\}$
$= -a \times \{bc - cb\} - b \times \{-ac - ca\} + c \times \{ab + ba\}$
$= -a \times 0 - b \times (-2ac) + c \times 2ab = 2abc + 2abc = 4abc.$ ∎

§ Problem 1.1.6. $\begin{vmatrix} a & h & g \\ h & b & f \\ g & f & e \end{vmatrix} = abc + 2fgh - af^2 - bg^2 - ch^2.$ ◊

§§ Solution. $\begin{vmatrix} a & h & g \\ h & b & f \\ g & f & e \end{vmatrix} = a \times \begin{vmatrix} b & f \\ f & c \end{vmatrix} - h \times \begin{vmatrix} h & f \\ g & c \end{vmatrix} + g \times \begin{vmatrix} h & b \\ g & f \end{vmatrix}$

$= a \times \{b \times c - f \times f\} - h \times \{h \times c - f \times g\} + g\{h \times f - b \times g\}$
$= a \times \{bc - f^2\} - h \times \{hc - fg\} + g\{hf - bg\}$
$= abc - af^2 - h^2c + fgh + fgh - bg^2 = abc + 2fgh - af^2 - bg^2 - ch^2.$ ∎

Chapter 2

Coordinates : Lengths of Straight Lines and Areas of Triangles

2.1 Coordinates

§ Problem 2.1.1. *To find the coordinates of the point which externally divides in a given ratio $(m_1 : m_2)$ the line joining two given points (x_1, y_1) and (x_2, y_2).* ◇

§§ Solution. Let P_1 be the point (x_1, y_1), P_2 the point (x_2, y_2), and Q the required point, so that we have

$$P_1Q : QP_2 :: m_1 : m_2.$$

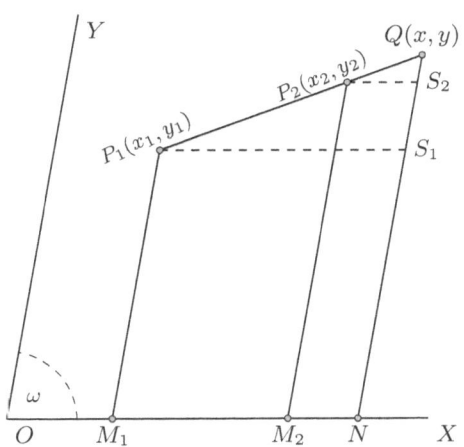

2.2. Length of Straight Line

Let Q be the point (x, y) so that if P_1M_1, P_2M_2 and QN be drawn parallel to the axis of y to meet the axis of x in M_1, M_2 and N, we have

$OM_1 = x_1$, $M_1P_1 = y_1$, $OM_2 = x_2$, $M_2P_2 = y_2$, $ON = x$, $NQ = y$.

Draw P_1S_1 and P_2S_2, parallel to OX, to meet QN in S_1 and S_2 respectively.

Then
$$P_1S_1 = M_1N = ON - OM_1 = x - x_1,$$
$$P_2S_2 = M_2N = ON - OM_2 = x - x_2,$$
$$QS_1 = QN - S_1N = QN - P_1M_1 = y - y_1,$$
and
$$QS_2 = QN - S_2N = QN - P_2M_2 = y - y_2.$$

From the similar triangles P_1S_1Q and P_2S_2Q we have
$$\frac{m_1}{m_2} = \frac{P_1Q}{QP_2} = \frac{P_1S_1}{P_2S_2} = \frac{x - x_1}{x - x_2}$$
$$\therefore m_1(x - x_2) = m_2(x - x_1),$$

i.e.
$$x = \frac{m_1 x_2 - m_2 x_1}{m_1 - m_2}.$$

Again
$$\frac{m_1}{m_2} = \frac{P_1Q}{QP_2} = \frac{S_1Q}{S_2Q} = \frac{y - y_1}{y - y_2},$$

so that
$$m_1(y - y_2) = m_2(y - y_1),$$

and hence
$$y = \frac{m_1 y_2 - m_2 y_1}{m_1 - m_2}.$$

The coordinates of the point which divides the line P_1P_2 externally in the given ratio $m_1 : m_2$, are therefore

$$\frac{m_1 x_2 - m_2 x_1}{m_1 - m_2} \text{ and } \frac{m_1 y_2 - m_2 y_1}{m_1 - m_2}. \qquad \blacksquare$$

2.2 Length of Straight Line

Find the distances between the following pairs of points

§ Problem 2.2.1. $(2, 3)$ *and* $(5, 7)$. ◊

§§ Solution.

Distance
$$= \sqrt{(2-5)^2 + (3-7)^2}$$
$$= \sqrt{9 + 16} = \sqrt{25}$$
$$= 5. \qquad \blacksquare$$

§ Problem 2.2.2. $(4, -7)$ *and* $(-1, 5)$. ◊

§§ Solution.

Distance
$$= \sqrt{(4-(-1))^2 + (-7-5)^2)}$$
$$= \sqrt{25 + 144} = \sqrt{169}$$
$$= 13. \qquad \blacksquare$$

§ Problem 2.2.3. $(-3, -2)$ *and* $(-6, 7)$, *the axes being inclined at* $60°$. ◊

2.2. Length of Straight Line

§§ Solution. Distance
$$= \sqrt{(-3-(-6))^2 + (-2-7)^2 + 2(-3-(-6))(-2-7)\cos 60°}$$
$$= \sqrt{9 + 9^2 + 2 \times 3 \times (-9) \times \frac{1}{2}}$$
$$= \sqrt{9 \times 7} = 3\sqrt{7}.$$
∎

§ Problem 2.2.4. $(a, 0)$ *and* $(0, b)$.   ◊

§§ Solution. Distance
$$= \sqrt{(a-0)^2 + (0-b)^2}$$
$$= \sqrt{a^2 + b^2}.$$
∎

§ Problem 2.2.5. $(b+c, c+a)$ *and* $(c+a, a+b)$.   ◊

§§ Solution. Distance
$$= \sqrt{(b+c-(c+a))^2 + (c+a-(a+b))^2} = \sqrt{(b-a)^2 + (c-b)^2}$$
$$= \sqrt{b^2 + a^2 - 2ab + c^2 + b^2 - 2bc}$$
$$= \sqrt{a^2 + 2b^2 + c^2 - 2ab - 2bc}.$$
∎

§ Problem 2.2.6. $(a\cos\alpha, a\sin\alpha)$ *and* $(a\cos\beta, a\sin\beta)$.   ◊

§§ Solution. Distance
$$= \sqrt{(a\cos\alpha - a\cos\beta)^2 + (a\sin\alpha - a\sin\beta)^2}$$
$$= a\sqrt{(\cos\alpha - \cos\beta)^2 + (\sin\alpha - \sin\beta)^2} \quad (2.1)$$

To find a simplified expressions for $(\cos\alpha - \cos\beta)$ and $(\sin\alpha - \sin\beta)$, let us recall that, for all values of a and b,
$$\sin(a+b) = \sin a \cos b + \cos a \sin b \quad (2.2)$$
$$\sin(a-b) = \sin a \cos b - \cos a \sin b \quad (2.3)$$

Subtracting (2.3) from (2.2), we get:
$$\sin(a+b) - \sin(a-b) = 2\cos a \sin b \quad (2.4)$$

Let us introduce the variables α and β, so that $a+b = \alpha$ and $a-b = \beta$, we have
$$a = \frac{\alpha+\beta}{2} \text{ and } b = \frac{\alpha-\beta}{2}$$

Hence (2.4) becomes
$$\sin\alpha - \sin\beta = 2\cos\frac{\alpha+\beta}{2}\sin\frac{\alpha-\beta}{2} \quad (2.5)$$

Similarly,
$$\cos(a+b) = \cos a \cos b - \sin a \sin b \quad (2.6)$$
$$\cos(a-b) = \cos a \cos b + \sin a \sin b \quad (2.7)$$

Subtracting these, we get :
$$\cos(a+b) - \cos(a-b) = -2\sin a \sin b \quad (2.8)$$

Substituting α and β, this becomes
$$\cos\alpha - \cos\beta = -2\sin\frac{\alpha+\beta}{2}\sin\frac{\alpha-\beta}{2} \quad (2.9)$$

With the help of (2.5) and (2.9), (2.1) becomes

Distance
$$= a\sqrt{\left(-2\sin\frac{\alpha+\beta}{2}\sin\frac{\alpha-\beta}{2}\right)^2 + \left(2\sin\frac{\alpha-\beta}{2}\cos\frac{\alpha+\beta}{2}\right)^2}$$

2.2. Length of Straight Line

$$= 2a\sqrt{\sin^2\frac{\alpha+\beta}{2}\,\sin^2\frac{\alpha-\beta}{2}+\sin^2\frac{\alpha-\beta}{2}\,\cos^2\frac{\alpha+\beta}{2}}$$

$$= 2a\sqrt{\sin^2\frac{\alpha-\beta}{2}\left(\sin^2\frac{\alpha+\beta}{2}+\cos^2\frac{\alpha+\beta}{2}\right)}$$

$$= 2a\sqrt{\sin^2\frac{\alpha-\beta}{2}(1)}$$

$$= 2a\sin\frac{\alpha-\beta}{2}.\qquad\blacksquare$$

§ Problem 2.2.7. $(am_1^2, 2am_1)$ *and* $(am_2^2, 2am_2)$. ◊
§§ Solution.
Distance
$$= \sqrt{(am_1^2 - am_2^2)^2 + (2am_1 - 2am_2)^2}$$
$$= a\sqrt{(m_1^2 - m_2^2)^2 + 4(m_1 - m_2)^2}$$
$$= a\sqrt{(m_1 + m_2)^2(m_1 - m_2)^2 + 4(m_1 - m_2)^2}$$
$$= a\sqrt{(m_1 - m_2)^2\left\{(m_1 + m_2)^2 + 4\right\}}$$
$$= a(m_1 - m_2)\sqrt{(m_1 + m_2)^2 + 4}.\qquad\blacksquare$$

§ Problem 2.2.8. *Lay down in a figure the positions of the points* $(1, -3)$ *and* $(-2, 1)$, *and prove that the distance between them is* 5. ◊
§§ Solution. Distance
$$= \sqrt{(1-(-2))^2 + (-3-1)^2}$$
$$= \sqrt{9+16} = \sqrt{25} = 5.$$

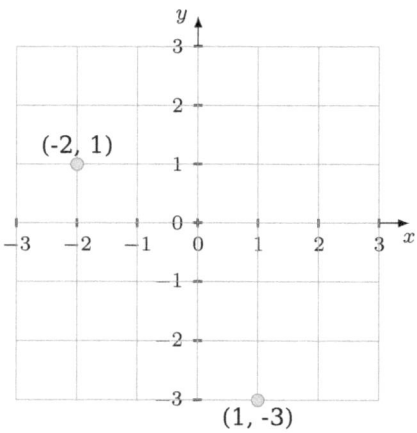

$\blacksquare$

§ Problem 2.2.9. *Find the value of* x_1 *if the distance between the points* $(x_1, 2)$ *and* $(3, 4)$ *be* 8. ◊
§§ Solution.
$$8 = \sqrt{(x_1 - 3)^2 + (2 - 4)^2}$$
$$8^2 = x_1^2 + 9 - 6x_1 + 4$$

2.2. Length of Straight Line

$$x_1^2 - 6x_1 - 51 = 0$$

Hence
$$x_1 = \frac{-(-6) \pm \sqrt{(-6)^2 - 4 \times 1 \times (-51)}}{2 \times 1}$$
$$= \frac{6 \pm \sqrt{36 + 204}}{2} = \frac{6 \pm \sqrt{240}}{2}$$
$$= 3 \pm \sqrt{60} = 3 \pm 2\sqrt{15}.$$
∎

§ Problem 2.2.10. *A line is of length 10 and one end is at the point $(2, -3)$; if the abscissa of the other end be 10, prove that its ordinate must be 3 or -9.* ◊

§§ Solution. Let the ordinate of second point be y,

then
$$10 = \sqrt{(2-10)^2 + (-3-y)^2}$$
$$10 = \sqrt{64 + 9 + y^2 + 6y}$$
$$\therefore y^2 + 6y - 27 = 0$$

$$\therefore y = \frac{-6 \pm \sqrt{6^2 - 4 \times 1 \times (-27)}}{2}$$
$$= \frac{-6 \pm \sqrt{36 + 108}}{2}$$
$$= \frac{-6 \pm \sqrt{144}}{2}$$
$$= -3 \pm \sqrt{36} = -3 \pm 6$$
$$= -3 + 6, \text{ or } -3 - 6 = 3, \text{ or } -9.$$
∎

§ Problem 2.2.11. *Prove that the points $(2a, 4a)$, $(2a, 6a)$, and $(2a + \sqrt{3}a, 5a)$ are the vertices of an equilateral triangle whose side is $2a$.* ◊

§§ Solution. Let the points be A, B and C respectively. Then
$$AB = \sqrt{(2a - 2a)^2 + (4a - 6a)^2} = \sqrt{(2a)^2} = 2a.$$
$$BC = \sqrt{(2a - (2a + \sqrt{3}a))^2 + (6a - 5a)^2} = \sqrt{3a^2 + a^2} = \sqrt{4a^2} = 2a.$$
$$CA = \sqrt{((2a + \sqrt{3}a) - 2a)^2 + (5a - 4a)^2} = \sqrt{3a^2 + a^2} = \sqrt{4a^2} = 2a.$$

$$\therefore AB = BC = CA = 2a.$$

Hence it is proved that these points are the vertices of an equilateral triangle whose side is $2a$. ∎

§ Problem 2.2.12. *Prove that the points $(-2, -1)$, $(1, 0)$, $(4, 3)$, and $(1, 2)$ are at the vertices of a parallelogram.* ◊

§§ Solution. Let the points be A, B, C and D respectively.
$$AB = \sqrt{(-2-1)^2 + (-1-0)^2} = \sqrt{9+1} = \sqrt{10}$$
$$BC = \sqrt{(1-4)^2 + (0-3)^2} = \sqrt{9+9} = 3\sqrt{2}$$
$$CD = \sqrt{(4-1)^2 + (3-2)^2} = \sqrt{9+1} = \sqrt{10}$$
$$AD = \sqrt{(1-(-2))^2 + (2-(-1))^2} = \sqrt{9+9} = 3\sqrt{2}$$

$$\therefore AB = CD \text{ and } BC = AD$$

2.2. Length of Straight Line

Since the opposite sides are equal, ∴ it is a parallelogram.

Alternative Solution.

The middle point of the diagonal AC is $\left(\dfrac{-2+4}{2}, \dfrac{-1+3}{2}\right) = (1,1)$.

The middle point of the diagonal BD is $\left(\dfrac{1+1}{2}, \dfrac{0+2}{2}\right) = (1,1)$.

Since the diagonals bisect each other, hence it is a parallelogram. ∎

§ Problem 2.2.13. *Prove that the points $(2,-2)$, $(8,4)$, $(5,7)$, and $(-1,1)$ are at the angular points of a rectangle.* ◊

§§ Solution. Let the points be A, B, C and D respectively.

$$AB = \sqrt{(2-8)^2 + (-2-4)^2} = \sqrt{36+36} = \sqrt{72} = 6\sqrt{2}.$$
$$BC = \sqrt{(8-5)^2 + (4-7)^2} = \sqrt{9+9} = \sqrt{18} = 3\sqrt{2}.$$
$$CD = \sqrt{(5-(-1))^2 + (7-1)^2} = \sqrt{36+36} = \sqrt{72} = 6\sqrt{2}.$$
$$AD = \sqrt{(-1-2)^2 + (1-(-2))^2} = \sqrt{9+9} = \sqrt{18} = 3\sqrt{2}.$$
$$AC = \sqrt{(2-5)^2 + (-2-7)^2} = \sqrt{9+81} = \sqrt{90} = 3\sqrt{10}.$$
$$BD = \sqrt{(8-(-1))^2 + (4-1)^2} = \sqrt{81+9} = \sqrt{90} = 3\sqrt{10}.$$
$$\therefore AB = CD,\ BC = AD$$
$$\text{and } AC = BD.$$

Since the opposite sides are equal as well as the diagonals, ∴ it is a rectangle.

Alternative Solution.

It is easy to see that $AC^2 = 90 = 72+18 = AB^2+BC^2 = AD^2+DC^2 = BD^2$, since the opposite sides are equal and the sides are at right angle to each other, hence it is a rectangle. ∎

§ Problem 2.2.14. *Prove that the point $\left(-\dfrac{1}{14}, \dfrac{39}{14}\right)$ is the center of the circle circumscribing the triangle whose angular points are $(1,1)$, $(2,3)$, and $(-2,2)$.* ◊

§§ Solution. Let the point $O(x,y)$ be the center of the circle and let the angular points of the triangle be A, B and C respectively.

It is easy to see that the radius of the circle $= OA = OB = OC$.

Then, $$OA^2 = OB^2$$
$$\therefore (x-1)^2 + (y-1)^2 = (x-2)^2 + (y-3)^2$$
$$\therefore x^2 + 1 - 2x + y^2 + 1 - 2y = x^2 + 4 - 4x + y^2 + 9 - 6y$$
$$\therefore 2x + 4y = 11 \tag{2.10}$$

Similarly, $$OB^2 = OC^2$$
$$\therefore (x-2)^2 + (y-3)^2 = (x-(-2))^2 + (y-2)^2$$
$$\therefore x^2 + 4 - 4x + y^2 + 9 - 6y = x^2 + 4 + 4x + y^2 + 4 - 4y$$
$$\therefore 8x + 2y = 5 \tag{2.11}$$

$2\times$ (2.11) $-$ (2.10) leads to $14x = -1$. $\therefore x = -\dfrac{1}{14}$.

Substituting the value of x in (2.10), we get $4y = 11 + \dfrac{2}{14}$. $\therefore y = \dfrac{39}{14}$.

2.2. Length of Straight Line

Hence the coordinates of the center of the circle is $\left(-\dfrac{1}{14}, \dfrac{39}{14}\right)$. ∎

Find the coordinates of the point which

§ Problem 2.2.15. *divides the line joining the points* $(1, 3)$ *and* $(2, 7)$ *in the ratio* $3 : 4$. ◊

§§ Solution.

Abscissa
$$= \frac{3 \times 2 + 4 \times 1}{3 + 4} = \frac{6 + 4}{7} = \frac{10}{7}.$$

Ordinate
$$= \frac{3 \times 7 + 4 \times 3}{3 + 4} = \frac{21 + 12}{7} = \frac{33}{7}.$$

Hence the point is $\left(\dfrac{10}{7}, \dfrac{33}{7}\right)$. ∎

§ Problem 2.2.16. *divides the same line in the ratio* $3 : -4$. ◊

§§ Solution.

Abscissa
$$= \frac{3 \times 2 + (-4) \times 1}{3 + (-4)} = \frac{6 - 4}{-1} = -2.$$

Ordinate
$$= \frac{3 \times 7 + (-4) \times 3}{3 + (-4)} = \frac{21 - 12}{-1} = -9.$$

Hence the point is $(-2, -9)$. ∎

§ Problem 2.2.17. *divides, internally and externally, the line joining* $(-1, 2)$ *to* $(4, -5)$ *in the ratio* $2 : 3$. ◊

§§ Solution. The coordinates of the point, which divides internally
$$= \left(\frac{2 \times 4 + 3 \times (-1)}{2 + 3}, \frac{2 \times (-5) + 3 \times 2}{2 + 3}\right)$$
$$= \left(1, -\frac{4}{5}\right).$$

The coordinates of the point, which divides externally
$$= \left(\frac{2 \times 4 - 3 \times (-1)}{2 - 3}, \frac{2 \times (-5) - 3 \times 2}{2 - 3}\right)$$
$$= (-11, 16). $$ ∎

§ Problem 2.2.18. *divides, internally and externally, the line joining* $(-3, -4)$ *to* $(-8, 7)$ *in the ratio* $7 : 5$. ◊

§§ Solution. The coordinates of the point, which divides internally
$$= \left(\frac{7 \times (-8) + 5 \times (-3)}{7 + 5}, \frac{7 \times 7 + 5 \times (-4)}{7 + 5}\right)$$
$$= \left(\frac{-56 - 15}{12}, \frac{49 - 20}{12}\right)$$
$$= \left(-\frac{71}{12}, \frac{29}{12}\right)$$
$$= \left(-5\frac{11}{12}, 2\frac{5}{12}\right).$$

The coordinates of the point, which divides externally
$$= \left(\frac{7 \times (-8) - 5 \times (-3)}{7 - 5}, \frac{7 \times 7 - 5 \times (-4)}{7 - 5}\right)$$

2.2. Length of Straight Line

$$= \left(\frac{-56+15}{2}, \frac{49+20}{2}\right)$$
$$= \left(-\frac{41}{2}, \frac{69}{12}\right)$$
$$= \left(-20\frac{1}{2}, 34\frac{1}{2}\right).$$

∎

§ Problem 2.2.19. *The line joining the points $(1,-2)$ and $(-3,4)$ is trisected; find the coordinates of the points of trisection.* ◊

§§ Solution. Let the points be A and B respectively. Let the points of trisection be P and Q respectively. Hence $AP : PB = 1 : 2$ and $AQ : QB = 2 : 1$

The coordinates of P

$$= \left(\frac{1\times(-3)+2\times 1}{1+2}, \frac{1\times 4+2\times(-2)}{1+2}\right) = \left(-\frac{1}{3}, 0\right).$$

The coordinates of Q

$$= \left(\frac{2\times(-3)+1\times 1}{2+1}, \frac{2\times 4+1\times(-2)}{2+1}\right) = \left(-\frac{5}{3}, 2\right).$$

∎

§ Problem 2.2.20. *The line joining the points $(-6,8)$ and $(8,-6)$ is divided into four equal parts; find the coordinates of the points of section.* ◊

§§ Solution. Let the points of the line be A and B respectively. Let the points of section be P, Q and R respectively. Hence $AP : PQ : QR : RB = 1 : 1 : 1 : 1$ and $AQ : QB = 1 : 1$.

It is not difficult to see that Q is the mid-point of AB.

∴ The coordinates of $Q = \left(\dfrac{-6+8}{2}, \dfrac{8+(-6)}{2}\right) = (1,1)$.

Similarly P is the mid-point of AQ.

∴ The coordinates of $P = \left(\dfrac{-6+1}{2}, \dfrac{8+1}{2}\right) = \left(-\dfrac{5}{2}, \dfrac{9}{2}\right)$.

Similarly R is the mid-point of QB.

∴ The coordinates of $R = \left(\dfrac{1+8}{2}, \dfrac{1+(-6)}{2}\right) = \left(\dfrac{9}{2}, -\dfrac{5}{2}\right)$.

Hence the points of section are $\left(-\dfrac{5}{2}, \dfrac{9}{2}\right)$; $(1,1)$; $\left(\dfrac{9}{2}, -\dfrac{5}{2}\right)$.

∎

§ Problem 2.2.21. *Find the coordinates of the points which divide, internally and externally, the line joining the point $(a+b, a-b)$ to the point $(a-b, a+b)$ in the ratio $a : b$.* ◊

§§ Solution. The coordinates of the point, which divides internally

$$= \left(\frac{a\times(a-b)+b\times(a+b)}{a+b}, \frac{a\times(a+b)+b\times(a-b)}{a+b}\right).$$
$$= \left(\frac{a^2+b^2}{a+b}, \frac{a^2+2ab-b^2}{a+b}\right).$$

The coordinates of the point, which divides externally

$$= \left(\frac{a\times(a-b)-b\times(a+b)}{a-b}, \frac{a\times(a+b)-b\times(a-b)}{a-b}\right).$$

2.2. Length of Straight Line

$$= \left(\frac{a^2 - 2ab - b^2}{a - b}, \frac{a^2 + b^2}{a - b}\right).$$ ∎

§ Problem 2.2.22. *The coordinates of the vertices of a triangle are (x_1, y_1), (x_2, y_2) and $x_3, y_3)$. The line joining the first two is divided in the ratio $l : k$, and the line joining this point of division to the opposite angular point is then divided in the ratio $m : k + l$. Find the coordinates of the latter point of section.* ◊

§§ Solution. The coordinates of the first point
$$= \left(\frac{lx_2 + kx_1}{l + k}, \frac{ly_2 + ky_1}{l + k}\right).$$

The coordinates of the latter point
$$= \left(\frac{mx_3 + (k + l)\dfrac{lx_2 + kx_1}{l + k}}{m + k + l}, \frac{my_3 + (k + l)\dfrac{ly_2 + ky_1}{l + k}}{m + k + l}\right).$$

$$= \left(\frac{kx_1 + lx_2 + mx_3}{k + l + m}, \frac{ky_1 + ly_2 + my_3}{k + l + m}\right).$$ ∎

§ Problem 2.2.23. *Prove that the coordinates, x and y, of the middle point of the line joining the point $(2, 3)$ to the point $(3, 4)$ satisfy the equation*
$$x - y + 1 = 0.$$ ◊

§§ Solution. The coordinates of the middle point
$$= \left(\frac{2+3}{2}, \frac{3+4}{2}\right)$$
$$= \left(\frac{5}{2}, \frac{7}{2}\right)$$

It is easy to see that $\dfrac{5}{2} - \dfrac{7}{2} + 1 = 0$. ∎

§ Problem 2.2.24. *If G be the centroid of a triangle ABC and O be any other point, prove that*
$$3\left(GA^2 + GB^2 + GC^2\right) = BC^2 + CA^2 + AB^2,$$
and $OA^2 + OB^2 + OC^2 = GA^2 + GB^2 + GC^2 + 3GO^2$. ◊

§§ Solution. Let the coordinates of A, B and C be (x_1, y_1), (x_2, y_2) and (x_3, y_3) respectively.

Hence the coordinates of the centroid G are given by
$$\left(\frac{x_1 + x_2 + x_3}{3}, \frac{y_1 + y_2 + y_3}{3}\right).$$

$$\begin{aligned}
GA^2 + GB^2 + GC^2 &= \left(\frac{x_1 + x_2 + x_3}{3} - x_1\right)^2 + \left(\frac{y_1 + y_2 + y_3}{3} - y_1\right)^2 \\
&+ \left(\frac{x_1 + x_2 + x_3}{3} - x_2\right)^2 + \left(\frac{y_1 + y_2 + y_3}{3} - y_2\right)^2 \\
&+ \left(\frac{x_1 + x_2 + x_3}{3} - x_3\right)^2 + \left(\frac{y_1 + y_2 + y_3}{3} - y_3\right)^2 \\
&= \left(\frac{x_2 + x_3 - 2x_1}{3}\right)^2 + \left(\frac{y_2 + y_3 - 2y_1}{3}\right)^2 \\
&+ \left(\frac{x_1 + x_3 - 2x_2}{3}\right)^2 + \left(\frac{y_1 + y_3 - 2y_2}{3}\right)^2
\end{aligned}$$

$$+ \left(\frac{x_1 + x_2 - 2x_3}{3}\right)^2 + \left(\frac{y_1 + y_2 - 2y_3}{3}\right)^2$$

$$= \left(\frac{x_2 + x_3 - 2x_1}{3}\right)^2 + \left(\frac{x_1 + x_3 - 2x_2}{3}\right)^2$$

$$+ \left(\frac{x_1 + x_2 - 2x_3}{3}\right)^2$$

$$+ \left(\frac{y_2 + y_3 - 2y_1}{3}\right)^2 + \left(\frac{y_1 + y_3 - 2y_2}{3}\right)^2$$

$$+ \left(\frac{y_1 + y_2 - 2y_3}{3}\right)^2$$

$$= \frac{(x_2 + x_3)^2 + 4x_1^2 - 4x_1(x_2 + x_3)}{9}$$

$$+ \frac{(x_1 + x_3)^2 + 4x_2^2 - 4x_2(x_1 + x_3)}{9}$$

$$+ \frac{(x_1 + x_2)^2 + 4x_3^2 - 4x_3(x_1 + x_2)}{9}$$

$$+ \frac{(y_2 + y_3)^2 + 4y_1^2 - 4y_1(y_2 + y_3)}{9}$$

$$+ \frac{(y_1 + y_3)^2 + 4y_2^2 - 4y_2(y_1 + y_3)}{9}$$

$$+ \frac{(y_1 + y_2)^2 + 4y_3^2 - 4y_3(y_1 + y_2)}{9}$$

$$= \frac{6x_1^2 + 6x_2^2 + 6x_3^2 - 6x_1x_2 - 6x_2x_3 - 6x_3x_1}{9}$$

$$+ \frac{6y_1^2 + 6y_2^2 + 6y_3^2 - 6y_1y_2 - 6y_2y_3 - 6y_3y_1}{9}$$

$$\therefore 3(GA^2 + GB^2 + GC^2) = 2x_1^2 + 2x_2^2 + 2x_3^2 - 2x_1x_2 - 2x_2x_3 - 2x_3x_1$$
$$+ 2y_1^2 + 2y_2^2 + 2y_3^2 - 2y_1y_2 - 2y_2y_3 - 2y_3y_1 \quad (2.12)$$

$$= (x_1^2 + x_2^2 - 2x_1x_2) + (x_2^2 + x_3^2 - 2x_2x_3)$$
$$+ (x_3^2 + x_1^2 - 2x_3x_1)$$
$$+ (y_1^2 + y_2^2 - 2y_1y_2) + (y_2^2 + y_3^2 - 2y_2y_3)$$
$$+ (y_3^2 + y_1^2 - 2y_3y_1)$$
$$= (x_1 - x_2)^2 + (x_2 - x_3)^2 + (x_3 - x_1)^2$$
$$+ (y_1 - y_2)^2 + (y_2 - y_3)^2 + (y_3 - y_1)^2$$
$$= AB^2 + BC^2 + CA^2.$$

Let (x, y) be the coordinates of O.
$\therefore 3GO^2$

$$= 3\left\{\left(\frac{x_1 + x_2 + x_3}{3} - x\right)^2 + \left(\frac{y_1 + y_2 + y_3}{3} - y\right)^2\right\}$$

$$= \frac{1}{3}\left\{(x_1 + x_2 + x_3 - 3x)^2 + (y_1 + y_2 + y_3 - 3y)^2\right\}$$

$$= \frac{1}{3}\left\{x_1^2 + x_2^2 + x_3^2 + 9x^2 + 2x_1x_2 + 2x_2x_3 + 2x_3x_1 - 6x(x_1 + x_2 + x_3)\right\}$$

2.2. Length of Straight Line

$$+ \frac{1}{3}\left\{y_1^2 + y_2^2 + y_3^2 + 9y^2 + 2y_1y_2 + 2y_2y_3 + 2y_3y_1 - 6y(y_1 + y_2 + y_3)\right\} \tag{2.13}$$

From (2.12) and (2.13):
$$\therefore GA^2 + GB^2 + GC^2 + 3GO^2$$
$$= \frac{1}{3}\left\{3(x_1^2 + x_2^2 + x_3^2) + 9x^2 - 6x(x_1 + x_2 + x_3)\right\}$$
$$+ \frac{1}{3}\left\{3(y_1^2 + y_2^2 + y_3^2) + 9y^2 - 6y(y_1 + y_2 + y_3)\right\}$$
$$= x_1^2 + x_2^2 + x_3^2 + 3x^2 - 2x(x_1 + x_2 + x_3)$$
$$+ y_1^2 + y_2^2 + y_3^2 + 3y^2 - 2y(y_1 + y_2 + y_3)$$
$$= (x - x_1)^2 + (x - x_2)^2 + (x - x_3)^2$$
$$+ (y - y_1)^2 + (y - y_2)^2 + (y - y_3)^2$$
$$= OA^2 + OB^2 + OC^2. \qquad \blacksquare$$

§ Problem 2.2.25. *Prove that the lines joining the middle points of opposite sides of a quadrilateral and the line joining the middle points of its diagonals meet in a point and bisect one another.* ◊

§§ Solution. Let the coordinates of the quadrilateral be $A(x_1, y_1)$, $B(x_2, y_2)$, $C(x_3, y_3)$ and $D(x_4, y_4)$ respectively.

The middle point of AB is $P\left(\dfrac{x_1 + x_2}{2}, \dfrac{y_1 + y_2}{2}\right)$.

The middle point of CD is $Q\left(\dfrac{x_3 + x_4}{2}, \dfrac{y_3 + y_4}{2}\right)$.

The middle point of the line PQ is
$$\left(\frac{x_1 + x_2 + x_3 + x_4}{4}, \frac{y_1 + y_2 + y_3 + y_4}{4}\right). \tag{2.14}$$

The middle point of BC is $R\left(\dfrac{x_2 + x_3}{2}, \dfrac{y_2 + y_3}{2}\right)$.

The middle point of AD is $S\left(\dfrac{x_1 + x_4}{2}, \dfrac{y_1 + y_4}{2}\right)$.

The middle point of the line RS is
$$\left(\frac{x_1 + x_2 + x_3 + x_4}{4}, \frac{y_1 + y_2 + y_3 + y_4}{4}\right). \tag{2.15}$$

The middle point of the diagonal AC is $U\left(\dfrac{x_1 + x_3}{2}, \dfrac{y_1 + y_3}{2}\right)$.

The middle point of the diagonal BD is $V\left(\dfrac{x_2 + x_4}{2}, \dfrac{y_2 + y_4}{2}\right)$.

The middle point of the line UV is
$$\left(\frac{x_1 + x_2 + x_3 + x_4}{4}, \frac{y_1 + y_2 + y_3 + y_4}{4}\right). \tag{2.16}$$

It is clear from the equations (2.14), (2.15) and (2.16) that the middle point of the lines PQ, RS and UV is the same. $\blacksquare$

§ Problem 2.2.26. *A, B, C, D are n points in a plane whose coordinates are (x_1, y_1), (x_2, y_2), (x_3, y_3), ... AB is bisected in the point G_1; G_1C is divided at G_2 in the ratio $1 : 2$; G_2D is divided at G_3 in the ratio $1 : 3$; G_3E at G_4 in the ratio $1 : 4$, and so on until all the points are exhausted. Show that the coordinates of the final point so obtained are*
$$\frac{x_1 + x_2 + x_3 + \ldots + x_n}{n} \text{ and } \frac{y_1 + y_2 + y_3 + \ldots + y_n}{n}.$$

*[This point is called the **Center of mean Position** of the n given points.]* ◊

2.2. Length of Straight Line

§§ Solution. Let us denote the abscissa of the final point G_{n-1} by $f(n-1)$. Then we have to prove that

$$f(n-1) = \frac{x_1 + x_2 + x_3 + \ldots + x_n}{n}. \tag{2.17}$$

We have to keep in mind that this corresponds to the ratio of the division being $1 : n-1$.

Let us adopt the principle of mathematical induction to prove it. Here, the base case is : $n = 2$.

In this case, (2.17) yields $f(1) = \dfrac{x_1 + x_2}{2}$.

The abscissa of G_1, i.e., $f(1)$ is the abscissa of the middle point of $A(x_1, y_1)$ and $B(x_2, y_2)$. Hence $f(1) = \dfrac{x_1 + x_2}{2}$.

Hence (2.17) hold true for the base case.

Let us assume that (2.17) holds true for a given natural number k, i.e.

$$f(k) = \frac{x_1 + x_2 + x_3 + \ldots + x_{k+1}}{k+1}. \tag{2.18}$$

We have to keep in mind that this corresponds to the ratio of the division being $1 : k$.

Now we have to see whether this equation holds true $k+1$. which corresponds to the ratio of the division being $1 : k+1$.

The abscissa of the ends of the line in question are
$$\frac{x_1 + x_2 + x_3 + \ldots + x_{k+1}}{k+1} \text{ and } x_{k+2}.$$

Hence the abscissa of the point which divides this line in the ratio $1 : k+1$ is

$$\frac{1 \times x_{k+2} + (k+1) \times \dfrac{x_1 + x_2 + x_3 + \ldots + x_{k+1}}{k+1}}{1 + (k+1)}.$$

Hence
$$f(k+1) = \frac{x_1 + x_2 + x_3 + \ldots + x_{k+1} + x_{k+2}}{k+2}. \tag{2.19}$$

Hence (2.17) hold true for the inductive step. Hence by mathematical induction, it holds true for any natural number n.

Similarly it holds true for the ordinate too.

Hence, it is proved that the coordinates of the final point G_{n-1} are
$$\frac{x_1 + x_2 + x_3 + \ldots + x_n}{n} \text{ and } \frac{y_1 + y_2 + y_3 + \ldots + y_n}{n}.$$

Alternative Solution :

Let us assume that the abscissa of the point G_{n-2} is
$$\frac{x_1 + x_2 + x_3 + \ldots + x_{n-1}}{n-1}.$$

Then the abscissa of the point G_{n-1} which divides the corresponding line with end abscissa being $\dfrac{x_1 + x_2 + x_3 + \ldots + x_{n-1}}{n-1}$ and x_n respectively in the ratio $1 : (n-1)$ is

$$\frac{1 \times x_n + (n-1) \times \dfrac{x_1 + x_2 + x_3 + \ldots + x_{n-1}}{n-1}}{1 + (n-1)}$$
$$= \frac{x_1 + x_2 + x_3 + \ldots + x_{n-1} + x_n}{n}.$$

2.2. Length of Straight Line

The abscissa of the point G_1, which bisects the line AB, is $\dfrac{x_1 + x_2}{2}$.

The abscissa of the point G_2, which divides the line G_1C in the ratio $1:2$, is

$$\dfrac{1 \times x_3 + 2 \times \dfrac{x_1 + x_2}{2}}{1+2}$$
$$= \dfrac{x_1 + x_2 + x_3}{3}.$$

Similar logic can be extended to the ordinate too.

Hence, by the mathematical induction, it is proved that the coordinates of the final point G_{n-1} are

$$\dfrac{x_1 + x_2 + x_3 + \ldots + x_n}{n} \text{ and } \dfrac{y_1 + y_2 + y_3 + \ldots + y_n}{n}.\qquad\blacksquare$$

§ Problem 2.2.27. *Prove that a point can be found which is at the same distance from each of the four points*

$$\left(am_1, \dfrac{a}{m_1}\right), \left(am_2, \dfrac{a}{m_2}\right), \left(am_3, \dfrac{a}{m_3}\right), \text{ and } \left(\dfrac{a}{m_1m_2m_3}, am_1m_2m_3\right). \diamond$$

§§ Solution. Let $O(x, y)$ be the point equidistant from A and D. Then $OA^2 = OD^2$.

$$\therefore (x - am_1)^2 + \left(y - \dfrac{a}{m_1}\right)^2 = \left(x - \dfrac{a}{m_1m_2m_3}\right)^2 + (y - am_1m_2m_3)^2$$

$$\therefore x^2 + a^2m_1^2 - 2axm_1 + y^2 + \dfrac{a^2}{m_1^2} - \dfrac{2ay}{m_1}$$

$$= x^2 + \dfrac{a^2}{m_1^2m_2^2m_3^2} - \dfrac{2ax}{m_1m_2m_3} + y^2 + a^2m_1^2m_2^2m_3^2 - 2aym_1m_2m_3$$

$$\therefore 2ax\left(\dfrac{1}{m_1m_2m_3} - m_1\right) + 2ay\left(m_1m_2m_3 - \dfrac{1}{m_1}\right)$$

$$= a^2\left(m_1^2m_2^2m_3^2 + \dfrac{1}{m_1^2m_2^2m_3^2} - m_1^2 - \dfrac{1}{m_1^2}\right)$$

$$\therefore 2x(1 - m_1^2m_2m_3) + 2y(m_1^2m_2^2m_3^2 - m_2m_3)$$

$$= a\left(m_1^3m_2^3m_3^3 + \dfrac{1}{m_1m_2m_3} - m_1^3m_2m_3 - \dfrac{m_2m_3}{m_1}\right)$$

$$\therefore (1 - m_1^2m_2m_3)(2x - 2ym_2m_3)$$

$$= \dfrac{a}{m_1m_2m_3}\left\{1 - m_2^2m_3^2 + m_1^4m_2^2m_3^2(m_2^2m_3^2 - 1)\right\}$$

$$= \dfrac{a(1 - m_2^3m_3^2)(1 - m_1^4m_2^2m_3^2)}{m_1m_2m_3}$$

$$\therefore (2x - 2ym_2m_3) = \dfrac{a(1 - m_2^3m_3^2)(1 + m_1^2m_2m_3)}{m_1m_2m_3}$$

$$\therefore (2x - 2ym_2m_3) = a\left\{\dfrac{1 + m_1^2m_2m_3 - m_2^2m_3^2 - m_1^2m_2^3m_3^3}{m_1m_2m_3}\right\}. \qquad (2.20)$$

Similarly, let $O(x, y)$ be the point equidistant from B and D. Then $OB^2 = OD^2$. Proceeding in the same way, we obtain

$$\therefore (2x - 2ym_3m_1) = a\left\{\dfrac{1 + m_1m_2^2m_3 - m_3^2m_1^2 - m_1^3m_2^2m_3^3}{m_1m_2m_3}\right\}. \qquad (2.21)$$

$$\therefore m_1 \times (2.20) - m_2 \times (2.21) \implies$$
$$2x(m_1 - m_2)$$

2.2. Length of Straight Line

$$= a\left\{\frac{1}{m_2 m_3} + m_1^2 - m_2 m_3 - m_1^2 m_2^2 m_3^2\right.$$
$$\left. - \frac{1}{m_1 m_3} - m_2^2 + m_1 m_3 + m_1^2 m_2^2 m_3^2\right\}$$
$$= a\left\{\frac{1}{m_3}\left(\frac{1}{m_2} - \frac{1}{m_1}\right) + (m_1^2 - m_2^2) + m_3(m_1 - m_2)\right\}$$
$$= a(m_1 - m_2)\left\{m_1 + m_2 + m_3 + \frac{1}{m_1 m_2 m_3}\right\}$$
$$\therefore 2x = a\left\{m_1 + m_2 + m_3 + \frac{1}{m_1 m_2 m_3}\right\} \quad (2.22)$$

Putting this in (2.21), we get

$$2y m_3 m_1 = a\left\{m_1 + m_2 + m_3 + \frac{1}{m_1 m_2 m_3} - \frac{1}{m_1 m_2 m_3} - m_2\right.$$
$$\left. + \frac{m_3 m_1}{m_2} + m_1^2 m_2 m_3^2\right\}$$
$$\therefore 2y = a\left\{\frac{1}{m_1} + \frac{1}{m_2} + \frac{1}{m_3} + m_1 m_2 m_3\right\} \quad (2.23)$$

Hence the coordinates of the point $O(x, y)$, which is equidistant from A, B and D, are

$$\frac{a}{2}\left(m_1 + m_2 + m_3 + \frac{1}{m_1 m_2 m_3}\right), \text{ and } \frac{a}{2}\left(\frac{1}{m_1} + \frac{1}{m_2} + \frac{1}{m_3} + m_1 m_2 m_3\right)$$

Looking at the symmetry of the coordinates, it is clear to see that it is also equidistant from B, C and D as well.

Hence, it is at the same distance from each of the four points A, B, C and D.

Alternative Solution:

Let $O(x, y)$ be the point equidistant from A, B and C.
Hence $OA^2 = OB^2$.

$$\therefore (x - am_1)^2 + \left(y - \frac{a}{m_1}\right)^2 = (x - am_2)^2 + \left(y - \frac{a}{m_2}\right)^2$$
$$\therefore (x - am_1)^2 - (x - am_2)^2 = \left(y - \frac{a}{m_2}\right)^2 - \left(y - \frac{a}{m_1}\right)^2$$
$$\therefore a(m_2 - m_1)\{2x - a(m_1 + m_2)\} = a\left(\frac{1}{m_1} - \frac{1}{m_2}\right)$$
$$\left\{2y - a\left(\frac{1}{m_1} + \frac{1}{m_2}\right)\right\}$$
$$\therefore 2x - a(m_1 + m_2) = \frac{2y}{m_1 m_2} - a(m_1 + m_2)\frac{1}{m_1^2 m_2^2}$$
$$\therefore 2x - \frac{2y}{m_1 m_2} = a(m_1 + m_2)\left(1 - \frac{1}{m_1^2 m_2^2}\right) \quad (2.24)$$

Similarly, $OB^2 = OC^2$ and from the symmetry,

$$\therefore 2x - \frac{2y}{m_2 m_3} = a(m_2 + m_3)\left(1 - \frac{1}{m_2^2 m_3^2}\right) \quad (2.25)$$

$(2.24) \times \dfrac{1}{m_3} - (2.25) \times \dfrac{1}{m_1} \implies$

$$2x\left(\frac{1}{m_3} - \frac{1}{m_1}\right)$$

2.2. Length of Straight Line

$$= \frac{a(m_1+m_2)}{m_3}\left(1-\frac{1}{m_1^2 m_2^2}\right) - \frac{a(m_2+m_3)}{m_1}\left(1-\frac{1}{m_2^2 m_3^2}\right)$$

$$= a\left(\frac{m_1}{m_3} + \frac{m_2}{m_3} - \frac{1}{m_1 m_2^2 m_3} - \frac{1}{m_1^2 m_2 m_3} - \frac{m_2}{m_1} - \frac{m_3}{m_1}\right.$$

$$\left. + \frac{1}{m_1 m_2 m_3^2} + \frac{1}{m_1 m_2^2 m_3}\right)$$

$$= a\left\{\frac{1}{m_1 m_2 m_3}\left(\frac{1}{m_3} - \frac{1}{m_1}\right) + m_2\left(\frac{1}{m_3} - \frac{1}{m_1}\right) + \frac{1}{m_1 m_3}(m_1^2 - m_3^2)\right\}$$

$$\therefore \frac{2x}{a}\left(\frac{m_1 - m_3}{m_1 m_3}\right)$$

$$= \frac{(m_1 - m_3)}{m_1 m_2 m_3 (m_1 m_3)} + (m_1 - m_3)\left(\frac{m_2}{m_1 m_3}\right) + \frac{(m_1 - m_3)(m_1 + m_3)}{m_1 m_3}$$

$$\therefore x = \frac{a}{2}\left(m_1 + m_2 + m_3 + \frac{1}{m_1 m_2 m_3}\right) \qquad (2.26)$$

Putting this value of x in (2.25), we get:

$$a\left(m_1 + m_2 + m_3 + \frac{1}{m_1 m_2 m_3} - m_2 - m_3 + \frac{1}{m_2 m_3^2} + \frac{1}{m_2^2 m_3}\right) = \frac{2y}{m_2 m_3}$$

$$\therefore y = \frac{a}{2}\left(\frac{1}{m_1} + \frac{1}{m_2} + \frac{1}{m_3} + m_1 m_2 m_3\right) \qquad (2.27)$$

Let us compute OA^2.

$$OA^2 = (x - am_1)^2 + \left(y - \frac{a}{m_1}\right)^2$$

$$= \frac{a^2}{4}\left[\left(m_2 + m_3 + \frac{1}{m_1 m_2 m_3} - m_1\right)^2 + \right.$$

$$\left. \left(\frac{1}{m_2} + \frac{1}{m_3} + m_1 m_2 m_3 - \frac{1}{m_1}\right)^2\right]$$

$$\therefore OA^2 = \frac{a^2}{4}\left[m_1^2 + m_2^2 + m_3^2 + \frac{1}{m_1^2} + \frac{1}{m_2^2} + \frac{1}{m_3^2} + m_1^2 m_2^2 m_3^2 + \frac{1}{m_1^2 m_2^2 m_3^2}\right] \qquad (2.28)$$

Let us compute OD^2.

$$OD^2 = \left(x - \frac{a}{m_1 m_2 m_3}\right)^2 + (y - am_1 m_2 m_3)^2$$

$$= \frac{a^2}{4}\left[\left(m_1 + m_2 + m_3 - \frac{1}{m_1 m_2 m_3}\right)^2 + \right.$$

$$\left. \left(\frac{1}{m_1} + \frac{1}{m_2} + \frac{1}{m_3} - m_1 m_2 m_3\right)^2\right]$$

$$\therefore OD^2 = \frac{c^2}{4}\left[m_1^2 + m_2^2 + m_3^2 + \frac{1}{m_1^2} + \frac{1}{m_2^2} + \frac{1}{m_3^2} + m_1^2 m_2^2 m_3^2 + \frac{1}{m_1^2 m_2^2 m_3^2}\right] \qquad (2.29)$$

From (2.28) and (2.29), $OA^2 = OD^2$.
Hence the point O is equidistant from A, B, C and D. ∎

2.3 Areas of Triangles

As per the Article 27, the area should be a positive quantity and to comply with this, the points A, B and C must be taken in the order in which they would be met by a person starting from A and walking round the triangle in such a manner that the *area of the triangle is always on his left hand*.

Find the areas of the triangles the coordinates of whose angular points are respectively

§ Problem 2.3.1. $(1,3)$, $(-7,6)$ *and* $(5,-1)$. ◊

§§ Solution. It is clear from the figure that the points should be taken in the order of A, B and C.

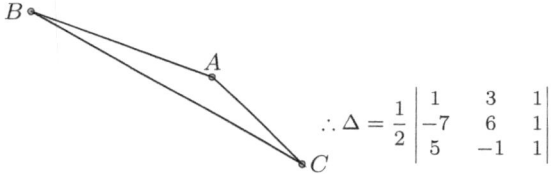

$$\therefore \Delta = \frac{1}{2} \begin{vmatrix} 1 & 3 & 1 \\ -7 & 6 & 1 \\ 5 & -1 & 1 \end{vmatrix}$$

$$\therefore \Delta = \frac{1}{2} \{1 \times (6 \times 1 - 1 \times (-1)) - 3 \times ((-7) \times 1 - 1 \times 5)$$
$$- 1 \times ((-7) \times (-1) - 5 \times 6)\}$$
$$= \frac{1}{2} \{6 + 1 + 3 \times (-12) + 7 - 30\} = \frac{1}{2} \{43 - 23\} = 10. \quad \blacksquare$$

§ Problem 2.3.2. $(0,4)$, $(3,6)$ *and* $(-8,-2)$. ◊

§§ Solution. It is clear from the figure that the points should be taken in the order of A, C and B.

$$\therefore \Delta = \frac{1}{2} \begin{vmatrix} 0 & 4 & 1 \\ -8 & -2 & 1 \\ 3 & 6 & 1 \end{vmatrix}$$

$$\therefore \Delta = \frac{1}{2} \{0 - 4 \times ((-8) \times 1 - 3 \times 1)$$
$$+ 1 \times ((-8) \times 6 - (-2) \times 3)\}$$
$$= \frac{1}{2} \{44 - 48 + 6\} = 1.$$

■

2.3. Areas of Triangles

§ Problem 2.3.3. $(5, 2)$, $(-9, -3)$ and $(-3, -5)$. ◊

§§ Solution. It is clear from the figure that the points should be taken in the order of A, B and C.

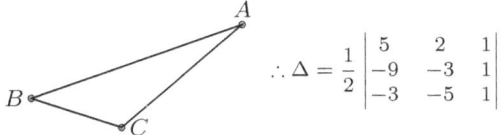

$$\therefore \Delta = \frac{1}{2} \begin{vmatrix} 5 & 2 & 1 \\ -9 & -3 & 1 \\ -3 & -5 & 1 \end{vmatrix}$$

$$\therefore \Delta = \frac{1}{2} \{5 \times (-3 \times 1 - 1 \times (-5)) - 2 \times ((-9) \times 1 - 1 \times (-3))$$
$$+ 1 \times ((-9) \times (-5) - (-3) \times (-3))\}$$
$$= \frac{1}{2} \{10 + 12 + 36\} = 29. \blacksquare$$

§ Problem 2.3.4. $(a, b+c)$, $(a, b-c)$ and $(-a, c)$. ◊

§§ Solution.
$$\Delta = \frac{1}{2} \begin{vmatrix} a & b+c & 1 \\ a & b-c & 1 \\ -a & c & 1 \end{vmatrix}$$
$$= \frac{1}{2} \{a(b - c - c) - (b+c)(a - (-a)) + (ac - (-a)(b-c))\}$$
$$= \frac{1}{2} \{ab - 2ac - 2ab - 2ac + ac + ab - ac\} = -2ac. \blacksquare$$

§ Problem 2.3.5. $(a, c+a)$, (a, c) and $(-a, c-a)$. ◊

§§ Solution.
$$\Delta = \frac{1}{2} \begin{vmatrix} a & c+a & 1 \\ a & c & 1 \\ -a & c-a & 1 \end{vmatrix}$$
$$= \frac{1}{2} \{a(c - (c-a)) - (c+a)(a - (-a)) + (a(c-a) - (-a)c)\}$$
$$= \frac{1}{2} \{a^2 - 2ac - 2a^2 + 2ac - a^2\} = -a^2. \blacksquare$$

§ Problem 2.3.6. $(a\cos\phi_1, b\sin\phi_1)$, $(a\cos\phi_2, b\sin\phi_2)$ and $(a\cos\phi_3, b\sin\phi_3)$. ◊

§§ Solution.
$$\Delta = \frac{1}{2} \begin{vmatrix} a\cos\phi_1 & b\sin\phi_1 & 1 \\ a\cos\phi_2 & b\sin\phi_2 & 1 \\ a\cos\phi_3 & b\sin\phi_3 & 1 \end{vmatrix}$$
$$= \frac{1}{2} \{a\cos\phi_1(b\sin\phi_2 - b\sin\phi_3) - b\sin\phi_1(a\cos\phi_2 - a\cos\phi_3)$$
$$+ (ab\cos\phi_2\sin\phi_3 - ab\cos\phi_3\sin\phi_2)\}$$
$$= \frac{ab}{2} \{(\sin\phi_2\cos\phi_1 - \sin\phi_1\cos\phi_2) + (\sin\phi_3\cos\phi_2 - \sin\phi_2\cos\phi_3)$$
$$+ (\sin\phi_1\cos\phi_3 - \sin\phi_3\cos\phi_1)\}$$
$$= \frac{ab}{2} \{\sin(\phi_2 - \phi_1) + \sin(\phi_3 - \phi_2) + \sin(\phi_1 - \phi_3)\}$$
$$= \frac{ab}{2} \left\{ 2\sin\frac{\phi_3 - \phi_1}{2}\cos\frac{2\phi_2 - \phi_1 - \phi_3}{2} + 2\sin\frac{\phi_1 - \phi_3}{2}\cos\frac{\phi_1 - \phi_3}{2} \right\}$$

2.3. Areas of Triangles

$$= ab \sin \frac{\phi_3 - \phi_1}{2} \left\{ \cos \frac{2\phi_2 - \phi_1 - \phi_3}{2} - \cos \frac{\phi_1 - \phi_3}{2} \right\}$$

$$= ab \sin \frac{\phi_3 - \phi_1}{2} 2 \sin \frac{1}{2} \left(\frac{2\phi_2 - \phi_1 - \phi_3}{2} + \frac{\phi_1 - \phi_3}{2} \right) \times$$

$$\sin \frac{1}{2} \left(\frac{\phi_1 - \phi_3}{2} - \frac{2\phi_2 - \phi_1 - \phi_3}{2} \right)$$

$$= 2ab \sin \frac{\phi_1 - \phi_2}{2} \sin \frac{\phi_2 - \phi_3}{2} \sin \frac{\phi_3 - \phi_1}{2}. \qquad \blacksquare$$

§ Problem 2.3.7. $(am_1^2, 2am_1)$, $(am_2^2, 2am_2)$ and $(am_3^2, 2am_3)$. $\Diamond$

§§ Solution.

$$\Delta = \frac{1}{2} \begin{vmatrix} am_1^2 & 2am_1 & 1 \\ am_2^2 & 2am_2 & 1 \\ am_3^2 & 2am_3 & 1 \end{vmatrix}$$

$$= \frac{1}{2} \left\{ am_1^2(2am_2 - 2am_3) - 2am_1(am_2^2 - am_3^2) \right.$$
$$\left. + (am_2^2 \times 2am_3 - am_3^2 \times 2am_2) \right\}$$

$$= a^2(m_2 - m_3)\left\{ m_1^2 - m_1(m_2 + m_3) + m_2 m_3 \right\}$$

$$= a^2(m_2 - m_3)\left\{ m_1(m_1 - m_2) - m_3(m_1 - m_2) \right\}$$

$$= -a^2(m_1 - m_2)(m_2 - m_3)(m_3 - m_1). \qquad \blacksquare$$

§ Problem 2.3.8. $\{am_1^2 m_2, a(m_1 + m_2)\}$, $\{am_2^2 m_3, a(m_2 + m_3)\}$ and $\{am_3^2 m_1, a(m_3 + m_1)\}$. $\Diamond$

§§ Solution.

$$\Delta = \frac{1}{2} \begin{vmatrix} am_1 m_2 & a(m_1 + m_2) & 1 \\ am_2 m_3 & a(m_2 + m_3) & 1 \\ am_3 m_1 & a(m_3 + m_1) & 1 \end{vmatrix}$$

$$= \frac{1}{2} \{ a^2 m_1 m_2 ((m_2 + m_3) - (m_3 + m_1)) - a^2 (m_1 + m_2)(m_2 m_3 - m_3 m_1)$$
$$+ a^2 (m_2 m_3 (m_3 + m_1) - m_3 m_1 (m_2 + m_3)) \}$$

$$= \frac{a^2}{2} \{ m_1 m_2 (m_2 - m_1) - (m_1 + m_2) m_3 (m_2 - m_1)$$
$$+ m_3 (m_2 m_3 + m_2 m_1 - m_1 m_2 - m_1 m_3) \}$$

$$= \frac{a^2(m_2 - m_1)}{2} \{ m_1 m_2 - m_1 m_3 - m_2 m_3 + m_3^2 \}$$

$$= \frac{a^2(m_2 - m_1)}{2} \{ m_2(m_1 - m_3) - m_3(m_1 - m_3) \}$$

$$= \frac{a^2}{2}(m_1 - m_2)(m_2 - m_3)(m_3 - m_1). \qquad \blacksquare$$

§ Problem 2.3.9. $\left\{ am_1, \dfrac{a}{m_1} \right\}$, $\left\{ am_2, \dfrac{a}{m_2} \right\}$ and $\left\{ am_3, \dfrac{a}{m_3} \right\}$. $\Diamond$

§§ Solution.

$$\Delta = \frac{1}{2} \begin{vmatrix} am_1 & \dfrac{a}{m_1} & 1 \\ am_2 & \dfrac{a}{m_2} & 1 \\ am_3 & \dfrac{a}{m_3} & 1 \end{vmatrix}$$

$$= \frac{1}{2} \left\{ am_1 \left(\frac{a}{m_2} - \frac{a}{m_3} \right) - \frac{a}{m_1}(am_2 - am_3) \right.$$

2.3. Areas of Triangles 21

$$
\begin{aligned}
&+ \left(am_2 \times \frac{a}{m_3} - am_3 \times \frac{a}{m_2} \right) \Big\} \\
&= \frac{a^2}{2} \left\{ m_1 \left(\frac{m_3 - m_2}{m_2 m_3} \right) - \left(\frac{m_2 - m_3}{m_1} \right) + \left(\frac{m_2^2 - m_3^2}{m_2 m_3} \right) \right\} \\
&= \frac{a^2(m_2 - m_3)}{2 m_1 m_2 m_3} \left\{ -m_1^2 - m_2 m_3 + m_1(m_2 + m_3) \right\} \\
&= \frac{a^2(m_2 - m_3)}{2 m_1 m_2 m_3} \left\{ m_1(m_3 - m_1) + m_2(m_1 - m_3) \right\} \\
&= \frac{a^2}{2 m_1 m_2 m_3} (m_1 - m_2)(m_2 - m_3)(m_3 - m_1).
\end{aligned}
$$
∎

Prove (by showing that the area of the triangle formed by them is *zero*) that the following sets of three points are in a straight line :

§ Problem 2.3.10. $(1, 4)$, $(3, -2)$ and $(-3, 16)$. ◊

§§ Solution.
$$
\begin{aligned}
\Delta &= \frac{1}{2} \begin{vmatrix} 1 & 4 & 1 \\ 3 & -2 & 1 \\ -3 & 16 & 1 \end{vmatrix} \\
&= \frac{1}{2} \{ (-2 - 16) - 4(3 + 3) + (48 - 6) \} \\
&= \frac{1}{2} \{ -18 - 24 + 42 \} = 0.
\end{aligned}
$$
∎

§ Problem 2.3.11. $\left(-\frac{1}{2}, 3 \right)$, $(-5, 6)$ and $(-8, 8)$. ◊

§§ Solution.
$$
\begin{aligned}
\Delta &= \frac{1}{2} \begin{vmatrix} -\frac{1}{2} & 3 & 1 \\ -5 & 6 & 1 \\ -8 & 8 & 1 \end{vmatrix} \\
&= \frac{1}{2} \left\{ -\frac{1}{2}(6 - 8) - 3(-5 + 8) + (-40 + 48) \right\} \\
&= \frac{1}{2} \{ 1 - 9 + 8 \} = 0.
\end{aligned}
$$
∎

§ Problem 2.3.12. $(a, b+c)$, $(b, c+a)$, and $(c, a+b)$. ◊

§§ Solution.
$$
\begin{aligned}
\Delta &= \frac{1}{2} \begin{vmatrix} a & b+c & 1 \\ b & c+a & 1 \\ c & a+b & 1 \end{vmatrix} \\
&= \frac{1}{2} \{ a(c + a - a - b) - (b + c)(b - c) + b(a + b) - c(c + a) \} \\
&= \frac{1}{2} \{ ac - ab - b^2 + c^2 + ab + b^2 - c^2 - ac \} = 0.
\end{aligned}
$$
∎

Find the areas of the quadrilaterals the coordinates of whose angular points, taken in order, are

§ Problem 2.3.13. $(1, 1)$, $(3, 4)$, $(5, -2)$ and $(4, -7)$. ◊

2.3. Areas of Triangles

§§ Solution.
$$\Delta = \frac{1}{2}\{(1 \times 4 - 3 \times 1) + (3 \times (-2) - 5 \times 4)$$
$$+ (5 \times (-7) - 4 \times (-2)) + (4 \times 1 - (-7) \times 1)\}$$
$$= \frac{1}{2}\{1 - 26 - 27 + 11\} = -\frac{41}{2} = -20\frac{1}{2}.$$

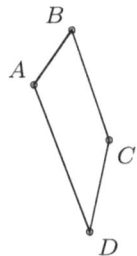

It is clear from the figure that the points should be taken in the order of $A(1,1)$, $D(4,-7)$, $C(5,-2)$ and $B(3,4)$ for the area to be a positive quantity.

$$\Delta = \frac{1}{2}\{(1(-7) - 4 \times 1) + (4(-2) - (-7)5) + (5 \times 4 - 3(-2))$$
$$+ (3 \times 1 - 1 \times 4)\}$$
$$= \frac{1}{2}\{-11 + 27 + 26 - 1\} = \frac{41}{2} = 20\frac{1}{2}. \qquad \blacksquare$$

§ Problem 2.3.14. $(-1, 6)$, $(-3, -9)$, $(5, -8)$ and $(3, 9)$. ◊
§§ Solution.
$$\Delta = \frac{1}{2}\{((-1)(-9) - 6(-3)) + ((-3)(-8) - (-9)5) + (5 \times 9 - (-8)3)$$
$$+ (3 \times 6 - 9(-1))\}$$
$$= \frac{1}{2}\{9 + 18 + 24 + 45 + 45 + 24 + 18 + 9\} = 96. \qquad \blacksquare$$

§ Problem 2.3.15. *If O be the origin, and if the coordinates of any two points P_1 and P_2 be respectively $(x-1, y_1)$ and (x_2, y_2), prove that*
$$OP_1 \cdot OP_2 \cdot \cos P_1 O P_2 = x_1 x_2 + y_1 y_2. \qquad ◊$$

§§ Solution. Let Δ be the area of the triangle $P_1 O P_2$.

$$\therefore \Delta = \frac{1}{2}\begin{vmatrix} 0 & 0 & 1 \\ x_1 & y_1 & 1 \\ x_2 & y_2 & 1 \end{vmatrix} = \frac{1}{2}(x_1 y_2 - x_2 y_1). \qquad (2.30)$$

We know that the area of a triangle is $\frac{1}{2}$ *base* $\times$ *height*.

$$\therefore \Delta = \frac{1}{2} OP_2 \cdot OP_1 \cdot \sin P_1 O P_2. \qquad (2.31)$$

From (2.30) and (2.31):
$$OP_1 \cdot OP_2 \cdot \sin P_1 O P_2 = x_1 y_2 - x_2 y_1. \qquad (2.32)$$

$$OP_1^2 \cdot OP_2^2 = \{(x_1 - 0)^2 + (y_1 - 0)^2\}^2 \{(x_2 - 0)^2 + (y_2 - 0)^2\}$$
$$= (x_1^2 + y_1^2)(x_2^2 + y_2^2). \qquad (2.33)$$

$(2.33) - (2.32)^2 \implies$
$$OP_1^2 \cdot OP_2^2 - OP_1^2 \cdot OP_2^2 \cdot \sin^2 P_1 O P_2 = (x_1^2 + y_1^2)(x_2^2 + y_2^2)$$
$$- (x_1 y_2 - x_2 y_1)^2$$

$$\therefore OP_1^2 \cdot OP_2^2 \cdot (1 - \sin^2 P_1OP_2) = x_1^2 x_2^2 + y_1^2 y_2^2 + 2x_1 x_2 y_1 y_2$$
$$\therefore OP_1^2 \cdot OP_2^2 \cdot \cos^2 P_1OP_2 = (x_1 x_2 + y_1 y_2)^2$$
$$\therefore OP_1 \cdot OP_2 \cdot \cos P_1OP_2 = x_1 x_2 + y_1 y_2.$$

Alternative Solution :
We know that
$$(P_1 P_2)^2 = (OP_1)^2 + (OP_2)^2 - 2 \cdot OP_1 \cdot OP_2 \cdot \cos P_1OP_2.$$

$$\therefore OP_1 \cdot OP_2 \cdot \cos P_1OP_2 = \frac{1}{2} \left[\left\{ (x_1^2 + y_1^2) + (x_2^2 + y_2^2) \right\} \right.$$
$$\left. - \left\{ (x_1 - x_2)^2 + (y_1 - y_2)^2 \right\} \right]$$
$$= x_1 x_2 + y_1 y_2. \qquad \blacksquare$$

2.4 Polar Coordinates

Lay down the positions of the points whose polar coordinates are

§ Problem 2.4.1. $(3, 45°)$. ◊

§§ Solution. :

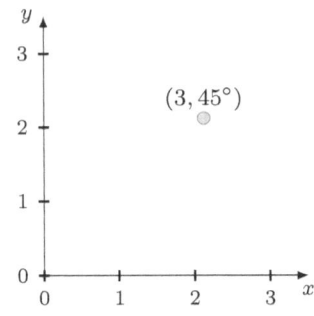

§ Problem 2.4.2. $(-2, -60°)$. ◊

§§ Solution. :

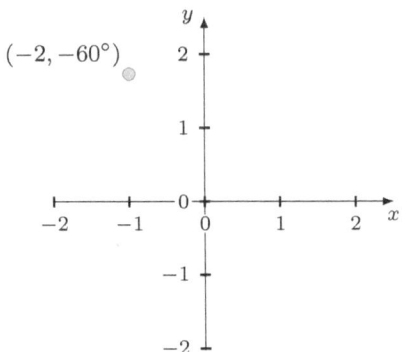

■

2.4. Polar Coordinates

§ Problem 2.4.3. $(4, 135°)$.

§§ Solution. :

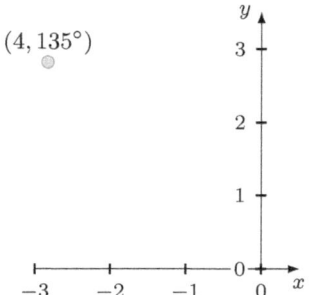

§ Problem 2.4.4. $(2, 330°)$.

§§ Solution. :

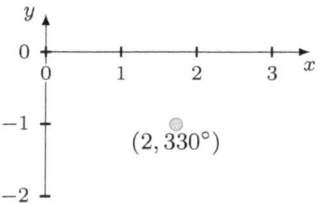

§ Problem 2.4.5. $(-1, 180°)$.

§§ Solution. :

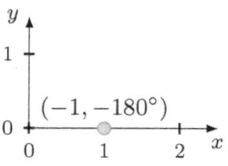

§ Problem 2.4.6. $(1, -210°)$.

§§ Solution. :

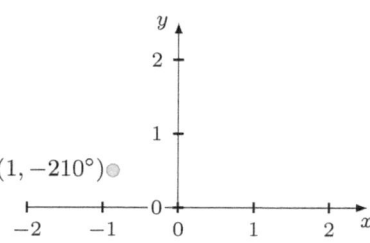

2.4. Polar Coordinates

§ Problem 2.4.7. $(5, -675°)$. ◊

§§ Solution. :

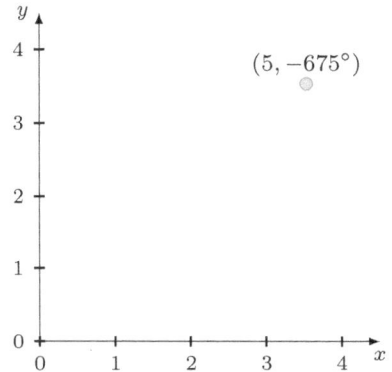

Let us assume that $a = 2$ for the sake of plotting the points in the following problems.

§ Problem 2.4.8. $\left(a, \dfrac{\pi}{2}\right)$. ◊

§§ Solution. :

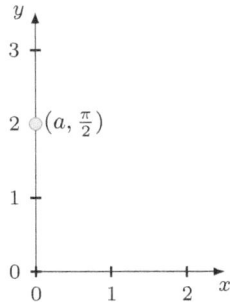

§ Problem 2.4.9. $\left(2a, -\dfrac{\pi}{2}\right)$. ◊

§§ Solution. :

2.4. Polar Coordinates

§ Problem 2.4.10. $\left(-a, \dfrac{\pi}{6}\right)$.

§§ Solution. :

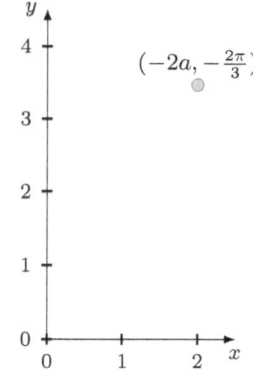

§ Problem 2.4.11. $\left(-2a, -\dfrac{2\pi}{3}\right)$.

§§ Solution. :

Let us denote the length by δ.
Find the lengths of the straight lines joining the pairs of points whose polar coordinates are

§ Problem 2.4.12. $(2, 30°)$ and $(4, 120°)$.

§§ Solution.
$$\delta^2 = 2^2 + 4^2 - 2 \cdot 2 \cdot 4\cos(30° - 120°).$$
$$= 4 + 16 - 16\cos(90°) = 20.$$
$$\therefore \delta = \sqrt{20} = 2\sqrt{5}.$$

§ Problem 2.4.13. $(-3, 45°)$ and $(7, 105°)$.

§§ Solution.
$$\delta^2 = (-3)^2 + 7^2 - 2 \cdot (-3) \cdot 7\cos(45° - 105°).$$
$$= 9 + 49 + 42\cos(60°) = 79.$$
$$\therefore \delta = \sqrt{79}.$$

2.4. Polar Coordinates

§ Problem 2.4.14. $\left(a, \dfrac{\pi}{2}\right)$ and $\left(3a, \dfrac{\pi}{6}\right)$. ◊

§§ Solution.
$$\delta^2 = a^2 + (3a)^2 - 2 \cdot a \cdot 3a \cos\left(\dfrac{\pi}{2} - \dfrac{\pi}{6}\right).$$
$$= 10a^2 - 6a^2 \cos\left(\dfrac{\pi}{3}\right) = 7a^2.$$
$$\therefore \delta = \sqrt{7}a. \qquad \blacksquare$$

§ Problem 2.4.15. *Prove that the points* $(0,0)$, $\left(3, \dfrac{\pi}{2}\right)$, *and* $\left(3, \dfrac{\pi}{6}\right)$ *form an equilateral triangle.* ◊

§§ Solution. Let us denote the points by $A(0,0)$, $B\left(3, \dfrac{\pi}{2}\right)$ and $C\left(3, \dfrac{\pi}{6}\right)$ respectively.

Let us compute the lengths of the sides of the triangle ABC.
$$AB^2 = 0^2 + 3^2 - 2 \cdot 0 \cdot 3 \cdot \cos\left(0 - \dfrac{\pi}{2}\right) = 9.$$
$$BC^2 = 3^2 + 3^2 - 2 \cdot 3 \cdot 3 \cdot \cos\left(\dfrac{\pi}{2} - \dfrac{\pi}{6}\right) = 9.$$
$$CA^2 = 3^2 + 0^2 - 2 \cdot 3 \cdot 0 \cdot \cos\left(\dfrac{\pi}{6} - 0\right) = 9.$$
$$\therefore AB^2 = BC^2 = CA^2$$
$$\therefore AB = BC = CA.$$
Hence, these are the sides of the equilateral triangle ABC. $\blacksquare$

Find the areas of the triangles the coordinates of whose angular points are

§ Problem 2.4.16. $(1, 30°)$, $(2, 60°)$ *and* $(3, 90°)$. ◊

§§ Solution.
$$\Delta = \dfrac{1}{2}\{1 \cdot 2 \cdot \sin(60° - 30°) + 2 \cdot 3 \cdot \sin(90° - 60°) + 3 \cdot 1 \cdot \sin(30° - 90°)\}$$
$$= \dfrac{1}{2}\{2\sin 30° + 6 \sin 30° - 3 \sin 60°\}$$
$$= \dfrac{1}{2}\left\{4 - 3 \cdot \dfrac{\sqrt{3}}{2}\right\} = \dfrac{8 - 3\sqrt{3}}{4}. \qquad \blacksquare$$

§ Problem 2.4.17. $(-3, -30°)$, $(5, 150°)$ *and* $(7, 210°)$. ◊

§§ Solution.
$$\Delta = \dfrac{1}{2}\{(-3) \cdot 5 \cdot \sin(150° - (-30°)) + 5 \cdot 7 \cdot \sin(210° - 150°)$$
$$+ 7 \cdot (-3) \cdot \sin(-30° - 210°)\}$$
$$= \dfrac{1}{2}\{-15 \sin 180° + 35 \sin 60° + 21 \sin 240°\}$$
$$= \dfrac{1}{2}\left\{35 \cdot \dfrac{\sqrt{3}}{2} - 21 \cdot \dfrac{\sqrt{3}}{2}\right\}$$
$$= \dfrac{7\sqrt{3}}{2}. \qquad \blacksquare$$

§ Problem 2.4.18. $\left(-a, \dfrac{\pi}{6}\right)$, $\left(a, \dfrac{\pi}{2}\right)$ and $\left(-2a, -\dfrac{2\pi}{3}\right)$. ◊

2.4. Polar Coordinates

§§ Solution.
$$\Delta = \frac{1}{2}\left\{(-a)\cdot a\cdot \sin\left(\frac{\pi}{2} - \frac{\pi}{6}\right) + a\cdot (-2a)\cdot \sin\left(-\frac{2\pi}{3} - \frac{\pi}{2}\right)\right.$$
$$\left. + (-2a)\cdot(-a)\cdot \sin\left(\frac{\pi}{6} - \left(-\frac{2\pi}{3}\right)\right)\right\}$$
$$= \frac{1}{2}\left\{-a^2 \sin\frac{\pi}{3} + 2a^2 \sin\frac{7\pi}{6} + 2a^2 \sin\frac{5\pi}{6}\right\}$$
$$= \frac{1}{2}\left\{-a^2 \cdot \frac{\sqrt{3}}{2} - 2a^2 \cdot \frac{1}{2} + 2a^2 \cdot \frac{1}{2}\right\}$$
$$= -\frac{1}{4}a^2\sqrt{3}.$$

Since area is a positive quantity, $\therefore \Delta = \frac{1}{4}a^2\sqrt{3}$. ∎

Find the polar coordinates (drawing the figure in each case) of the points

§ Problem 2.4.19. $x = \sqrt{3}, y = 1$. ◊

§§ Solution. :

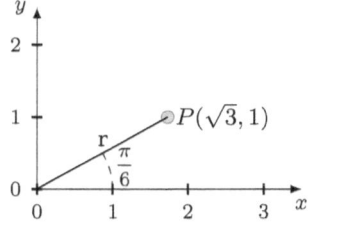

$r^2 = x^2 + y^2 = 3 + 1 = 4.$
$\therefore r = 2.$
$\theta = \tan^{-1}\frac{y}{x} = \tan^{-1}\frac{1}{\sqrt{3}}$
$\therefore \theta = \frac{\pi}{6}.$

∎

§ Problem 2.4.20. $x = -\sqrt{3}, y = 1$. ◊

§§ Solution. :

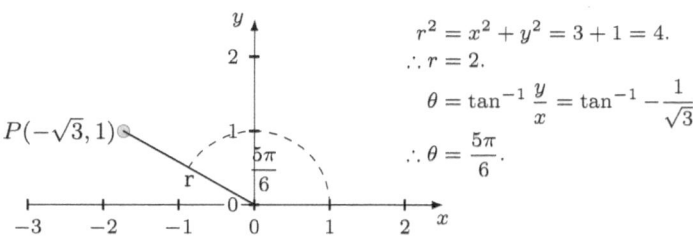

$r^2 = x^2 + y^2 = 3 + 1 = 4.$
$\therefore r = 2.$
$\theta = \tan^{-1}\frac{y}{x} = \tan^{-1}-\frac{1}{\sqrt{3}}$
$\therefore \theta = \frac{5\pi}{6}.$

∎

§ Problem 2.4.21. $x = -1, y = 1$. ◊

§§ Solution.
$$r^2 = x^2 + y^2 = 1 + 1 = 2.$$
$$\therefore r = \sqrt{2}.$$
$$\theta = \tan^{-1}\frac{y}{x} = \tan^{-1}-\frac{1}{\sqrt{1}}$$
$$\therefore \theta = \frac{3\pi}{4}.$$

2.4. Polar Coordinates

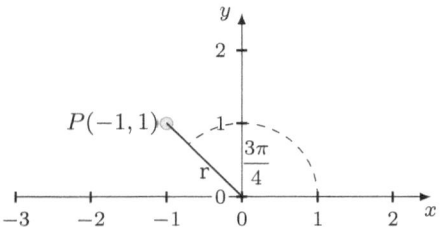

Find the Cartesian coordinates (drawing a figure in each case) of the points whose polar coordinates are

§ **Problem 2.4.22.** $\left(5, \dfrac{\pi}{4}\right)$.

§§ **Solution.**
$$x = r\cos\theta = 5\cos\frac{\pi}{4} = \frac{5}{\sqrt{2}}.$$
$$y = r\sin\theta = 5\sin\frac{\pi}{4} = \frac{5}{\sqrt{2}}.$$

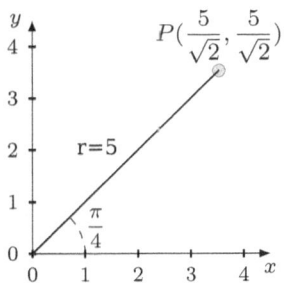

∴ the Cartesian coordinates are $\left(\dfrac{5}{\sqrt{2}}, \dfrac{5}{\sqrt{2}}\right)$.

§ **Problem 2.4.23.** $\left(-5, \dfrac{\pi}{3}\right)$.

§§ **Solution.**
$$x = r\cos\theta = -5\cos\frac{\pi}{3} = -\frac{5}{\sqrt{2}}.$$
$$y = r\sin\theta = -5\sin\frac{\pi}{3} = -\frac{5\sqrt{3}}{\sqrt{2}}.$$

∴ the Cartesian coordinates are $\left(-\dfrac{5}{\sqrt{2}}, -\dfrac{5\sqrt{3}}{\sqrt{2}}\right)$.

2.4. Polar Coordinates

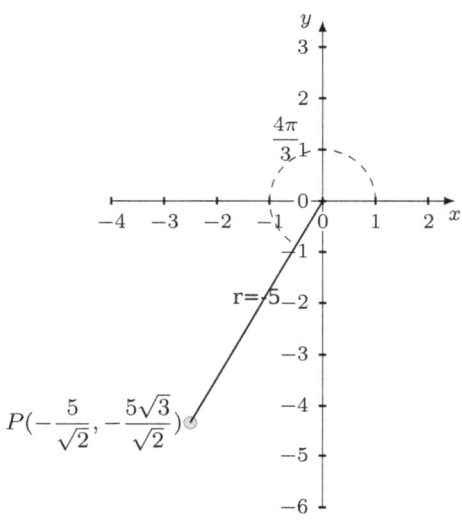

§ Problem 2.4.24. $\left(5, -\dfrac{\pi}{4}\right)$. ◊

§§ Solution.
$$x = r\cos\theta = 5\cos\left(-\dfrac{\pi}{4}\right) = \dfrac{5}{\sqrt{2}}.$$
$$y = r\sin\theta = 5\sin\left(-\dfrac{\pi}{4}\right) = -\dfrac{5}{\sqrt{2}}.$$
∴ the Cartesian coordinates are $\left(\dfrac{5}{\sqrt{2}}, -\dfrac{5}{\sqrt{2}}\right)$.

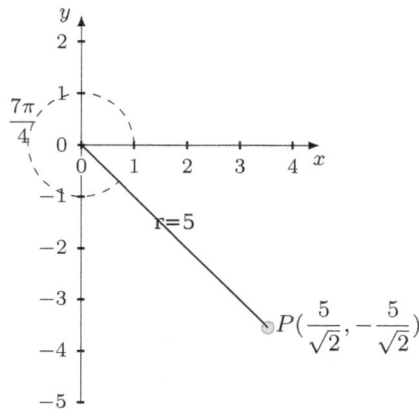

Change to polar coordinates the equations
§ Problem 2.4.25. $x^2 + y^2 = a^2$. ◊

2.4. Polar Coordinates

§§ Solution.
$$x^2 + y^2 = a^2$$
$$\therefore r^2 = a^2.$$
∎

§ Problem 2.4.26. $y = x \tan \alpha$. ◊
§§ Solution.
$$y = x \tan \alpha$$
$$\therefore r \sin \theta = r \cos \theta \cdot \tan \alpha$$
$$\therefore \tan \theta = \tan \alpha$$
$$\therefore \theta = \alpha.$$
∎

§ Problem 2.4.27. $x^2 + y^2 = 2ax$. ◊
§§ Solution.
$$x^2 + y^2 = 2ax$$
$$\therefore r^2 = 2ar \cos \theta$$
$$\therefore r = 2a \cos \theta.$$
∎

§ Problem 2.4.28. $x^2 - y^2 = 2ay$. ◊
§§ Solution.
$$x^2 - y^2 = 2ay$$
$$\therefore r^2 \cos^2 \theta - r^2 \sin^2 \theta = 2ar \sin \theta$$
$$\therefore r^2 (\cos^2 \theta - \sin^2 \theta) = 2ar \sin \theta$$
$$\therefore r \cos 2\theta = 2a \sin \theta.$$
∎

§ Problem 2.4.29. $x^3 = y^2(2a - x)$. ◊
§§ Solution.
$$x^3 = y^2(2a - x)$$
$$\therefore r^3 \cos^3 \theta = r^2 \sin^2 \theta (2a - r \cos \theta)$$
$$\therefore r \cos^3 \theta = (1 - \cos^2 \theta)(2a - r \cos \theta)$$
$$\therefore r \cos^3 \theta = 2a - r \cos \theta - 2a \cos^2 \theta + r \cos^3 \theta$$
$$\therefore r \cos \theta = 2a(1 - \cos^2 \theta)$$
$$\therefore r \cos \theta = 2a \sin^2 \theta.$$

Alternative Solution :
$$x^3 = y^2(2a - x)$$
$$\therefore r^3 \cos^3 \theta = r^2 \sin^2 \theta (2a - r \cos \theta)$$
$$\therefore r \cos^3 \theta = 2a \sin^2 \theta - r \sin^2 \theta \cos \theta$$
$$\therefore r \cos^3 \theta + r \sin^2 \theta \cos \theta) = 2a \sin^2 \theta$$
$$\therefore r \cos \theta \left(\cos^2 \theta + \sin^2 \theta \right) = 2a \sin^2 \theta$$
$$\therefore r \cos \theta = 2a \sin^2 \theta.$$
∎

§ Problem 2.4.30. $\left(x^2 + y^2 \right)^2 = a^2 \left(x^2 - y^2 \right)$. ◊
§§ Solution.
$$\left(x^2 + y^2 \right)^2 = a^2 \left(x^2 - y^2 \right)$$
$$\therefore r^4 = a^2 r^2 \left(\cos^2 \theta - \sin^2 \theta \right) = a^2 r^2 \cos 2\theta$$
$$\therefore r^2 = a^2 \cos 2\theta.$$
∎

Transform to Cartesian coordinates the equations

2.4. Polar Coordinates

§ Problem 2.4.31. $r = a$. ◊
§§ Solution.
$$r = a$$
$$\therefore r^2 = a^2$$
$$\therefore x^2 + y^2 = a^2.$$ ∎

§ Problem 2.4.32. $\theta = \tan^{-1} m$. ◊
§§ Solution.
$$\theta = \tan^{-1} m$$
$$\therefore \tan \theta = m$$
$$\therefore \frac{y}{x} = m$$
$$\therefore y = mx.$$ ∎

§ Problem 2.4.33. $r = a \cos \theta$. ◊
§§ Solution.
$$r = a \cos \theta$$
$$\therefore r^2 = ar \cos \theta$$
$$\therefore x^2 + y^2 = ax.$$ ∎

§ Problem 2.4.34. $r = a \sin 2\theta$. ◊
§§ Solution.
$$r = a \sin 2\theta$$
$$\therefore r = 2a \sin \theta \cos \theta$$
$$\therefore r^3 = 2ar^2 \sin \theta \cos \theta$$
$$\therefore (r^2)^{\frac{3}{2}} = 2a \cdot r \cos \theta \cdot r \sin \theta$$
$$\therefore (x^2 + y^2)^{\frac{3}{2}} = 2axy$$
$$\therefore (x^2 + y^2)^3 = 4a^2 x^2 y^2.$$ ∎

§ Problem 2.4.35. $r^2 = a^2 \cos 2\theta$. ◊
§§ Solution.
$$r^2 = a^2 \cos 2\theta$$
$$\therefore r^2 = a^2 \left(\cos^2 \theta - \sin^2 \theta \right)$$
$$\therefore r^4 = a^2 \left(r^2 cos^2 \theta - r^2 \sin^2 \theta \right)$$
$$\therefore \left(x^2 + y^2 \right)^2 = a^2 \left(x^2 - y^2 \right).$$ ∎

§ Problem 2.4.36. $r^2 \sin 2\theta = 2a^2$. ◊
§§ Solution.
$$r^2 \sin 2\theta = 2a^2$$
$$\therefore r^2 \cdot 2 \sin \theta \cdot \cos \theta = 2a^2$$
$$\therefore r \cos \theta \cdot r \sin \theta = a^2$$
$$\therefore xy = a^2.$$ ∎

§ Problem 2.4.37. $r^2 \cos 2\theta = a^2$. ◊

2.4. Polar Coordinates

§§ **Solution**.
$$r^2 \cos 2\theta = a^2$$
$$\therefore r^2 \left(\cos^2 \theta - \sin^2 \theta\right) = a^2$$
$$\therefore r^2 \cos^2 \theta - r^2 \sin^2 \theta = a^2$$
$$\therefore x^2 - y^2 = a^2.$$
∎

§ **Problem 2.4.38.** $r^{\frac{1}{2}} \cos \dfrac{\theta}{2} = a^{\frac{1}{2}}$. ◇

§§ **Solution**.
$$r^{\frac{1}{2}} \cos \frac{\theta}{2} = a^{\frac{1}{2}}$$
$$\therefore r \cos^2 \frac{\theta}{2} = a$$
$$\therefore r \cdot 2 \cos^2 \frac{\theta}{2} = 2a$$
$$\therefore r(1 + \cos \theta) = 2a$$
$$\therefore r = 2a - r \cos \theta$$
$$\therefore \sqrt{x^2 + y^2} = 2a - x$$
$$\therefore x^2 + y^2 = (2a - x)^2 = 4a^2 + x^2 - 4ax$$
$$\therefore y^2 + 4ax = 4a^2.$$
∎

§ **Problem 2.4.39.** $r^{\frac{1}{2}} = a^{\frac{1}{2}} \sin \frac{\theta}{2}$. ◇

§§ **Solution**.
$$r^{\frac{1}{2}} = a^{\frac{1}{2}} \sin \frac{\theta}{2}$$
$$\therefore 2r = a \cdot 2 \sin^2 \frac{\theta}{2}$$
$$\therefore 2r = a(1 - \cos \theta)$$
$$\therefore 2r^2 = ar - ar \cos \theta$$
$$\therefore 2(x^2 + y^2) = a\sqrt{(x^2 + y^2)} - ax$$
$$\therefore \left(2(x^2 + y^2) + ax\right)^2 = a^2(x^2 + y^2)$$
$$\therefore 4(x^2 + y^2)^2 + a^2 x^2 + 4ax(x^2 + y^2) = a^2 x^2 + a^2 y^2$$
$$\therefore 4(x^2 + y^2)^2 + 4ax(x^2 + y^2) = a^2 y^2$$
$$\therefore 4(x^2 + y^2)(x^2 + y^2 + ax) = a^2 y^2.$$
∎

§ **Problem 2.4.40.** $r(\cos 3\theta + \sin 3\theta) = 5k \sin \theta \cos \theta$. ◇

§§ **Solution**. Given equation is :
$$r(\cos 3\theta + \sin 3\theta) = 5k \sin \theta \cos \theta \tag{2.34}$$

Let us compute $\cos 3\theta$ as follows :
$\cos 3\theta = \cos(2\theta + \theta)$
$\quad = \cos 2\theta \cos \theta - \sin 2\theta \sin \theta$
$\quad = \cos(\theta + \theta) \cos \theta - \sin(\theta + \theta) \sin \theta$
$\quad = (\cos \theta \cos \theta - \sin \theta \sin \theta) \cos \theta - (\sin \theta \cos \theta + \cos \theta \sin \theta) \sin \theta$
$\quad = \cos^3 \theta - 3 \cos \theta \sin^2 \theta$

$$\therefore \cos 3\theta = \cos^3 \theta - 3 \cos \theta \sin^2 \theta \tag{2.35}$$

Similarly $\quad \sin 3\theta = 3 \sin \theta \cos^2 \theta - \sin^3 \theta \tag{2.36}$

2.4. Polar Coordinates

Putting these values into (2.34), we get:
$$r\left(\cos^3\theta - 3\cos\theta\sin^2\theta + 3\sin\theta\cos^2\theta - \sin^3\theta\right) = 5k\sin\theta\cos\theta$$
$$\therefore r^3\left(\cos^3\theta - 3\cos\theta\sin^2\theta + 3\sin\theta\cos^2\theta - \sin^3\theta\right) = 5kr^2\sin\theta\cos\theta$$
$$\therefore x^3 - 3xy^2 + 3x^2y - y^3 = 5kxy.$$

Alternative Solution :

$$\begin{aligned}\cos 3\theta &= \cos(2\theta + \theta) \\ &= \cos 2\theta \cos\theta - \sin 2\theta \sin\theta \\ &= \left[2\cos^2\theta - 1\right]\cos\theta - [2\sin\theta\cos\theta]\sin\theta \\ &= 2\cos^3\theta - \cos\theta - 2\sin^2\theta\cos\theta \\ &= 2\cos^3\theta - \cos\theta - 2\cos\theta(1-\cos^2)\theta \\ &= 4\cos^3\theta - 3\cos\theta\end{aligned}$$

$$\therefore \cos 3\theta = 4\cos^3\theta - 3\cos\theta \tag{2.37}$$

Similarly $\qquad \sin 3\theta = 3\sin\theta - 4\sin^3\theta \tag{2.38}$

Putting these values into (2.34), we get:
$$r(4\cos^3\theta - 3\cos\theta + 3\sin\theta - 4\sin^3\theta) = 5k\sin\theta\cos\theta$$
$$\therefore 4r(\cos^3\theta - \sin^3\theta) - 3r(\cos\theta - \sin\theta) = 5k\sin\theta\cos\theta$$
$$\therefore 4(r^3\cos^3\theta - r^3\sin^3\theta) - 3r^2(r\cos\theta - r\sin\theta) = 5k\cdot r\sin\theta\cdot r\cos\theta$$
$$\therefore 4(x^3 - y^3) - 3(x^2 + y^2)(x - y) = 5kxy$$
$$\therefore x^3 - 3xy^2 + 3x^2y - y^3 = 5kxy. \qquad\blacksquare$$

Chapter 3

Locus : Equation to a Locus

By taking a number of solutions, as in Arts. $39-41$**, sketch the loci of the following equations :**

§ **Problem 3.0.1.** $2x + 3y = 10$. ◊

§§ **Solution**.
$$2x + 3y = 10$$
$$\therefore y = \frac{10 - 2x}{3}$$

The following points satisfy this equation:
$P_1\left(0, \frac{10}{3}\right)$, $P_2\left(1, \frac{8}{3}\right)$, $P_3\left(2, 2\right)$, $P_4\left(5, 0\right)$, $P_5\left(-1, 4\right)$

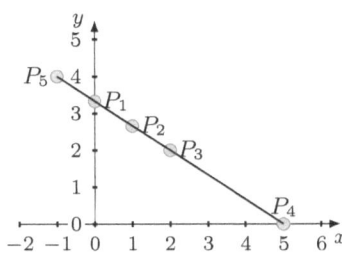

It is clear from the sketch that the locus is a straight line. ∎

§ **Problem 3.0.2.** $4x - y = 7$. ◊

§§ **Solution**.
$$4x - y = 7$$
$$\therefore y = 4x - 7.$$

Locus : Equation to a Locus

The following points satisfy this equation:
$P_1 (0, -7)$, $P_2 (1, -3)$, $P_3 (2, 1)$, $P_4 (3, 5)$.

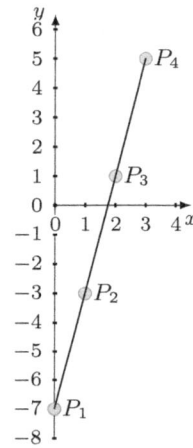

It is clear from the sketch that the locus is a straight line. ∎

§ Problem 3.0.3. $x^2 - 2ax + y^2 = 0$. ◊

§§ Solution.
$$x^2 - 2ax + y^2 = 0$$
$$\therefore (x-a)^2 + y^2 = a^2$$
$$\therefore y^2 = a^2 - (x-a)^2.$$

The following points satisfy this equation:
$$P_1(0,0),\ P_2(a, \pm a),\ P_3(2a, 0),\ P_4\left(\frac{3a}{2}, \pm\frac{\sqrt{3}a}{2}\right)$$

Let us assume $a = 1$.

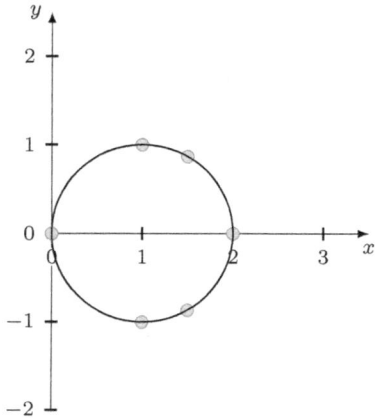

Locus : Equation to a Locus

It is clear from the sketch that the locus is a circle. ∎

§ **Problem 3.0.4.** $x^2 - 4ax + y^2 + 3a^2 = 0$. ◊

§§ **Solution**.
$$x^2 - 4ax + y^2 + 3a^2 = 0$$
$$\therefore (x - 2a)^2 + y^2 = a^2$$
$$\therefore y^2 = a^2 - (x - 2a)^2.$$

The following points satisfy this equation:
$$P_1(a, 0), \ P_2(2a, \pm a), \ P_3(3a, 0)$$

Let us assume $a = 1$.

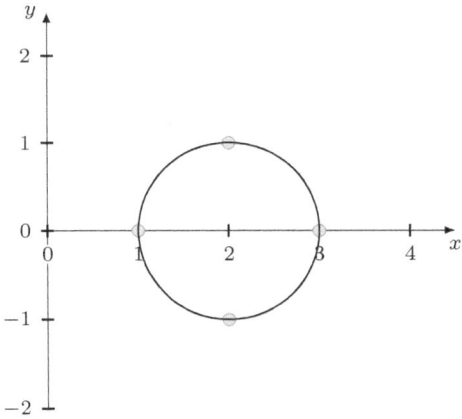

It is clear from the sketch that the locus is a circle. ∎

§ **Problem 3.0.5.** $y^2 = x$. ◊

§§ **Solution**.
$$y^2 = x$$
$$\therefore y = \pm\sqrt{x}.$$

The following points satisfy this equation:
$$P_1(0, 0), \ P_2(1, \pm 1), \ P_3(4, \pm 2), \ P_4(9, \pm 3)$$

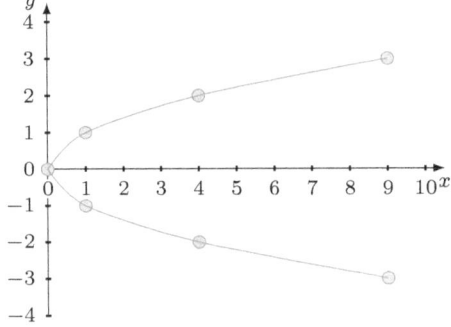

Locus : Equation to a Locus

It is clear from the sketch that the locus is a parabola. ∎

§ Problem 3.0.6. $3x = y^2 - 9$. ◊

§§ Solution.
$$3x = y^2 - 9$$
$$\therefore y = \pm\sqrt{3x + 9}.$$

The following points satisfy this equation:
$$P_1(0, \pm 3), \ P_2(-3, 0), \ P_3(9, \pm 6)$$

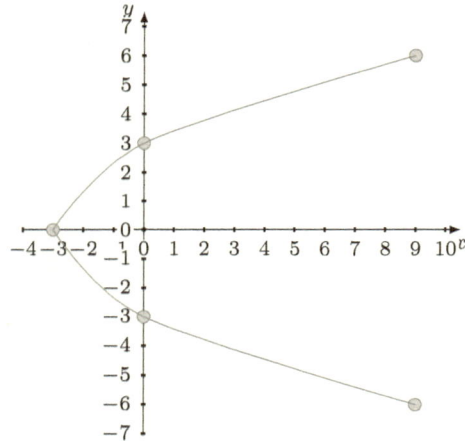

It is clear from the sketch that the locus is a parabola. ∎

§ Problem 3.0.7. $\dfrac{x^2}{4} + \dfrac{y^2}{9} = 1$. ◊

§§ Solution.
$$\frac{x^2}{4} + \frac{y^2}{9} = 1$$
$$\therefore y = \pm\sqrt{1 - \frac{x^2}{4}}.$$

The following points satisfy this equation:
$$P_1(0, \pm 3), \ P_2(\pm 2, 0)$$

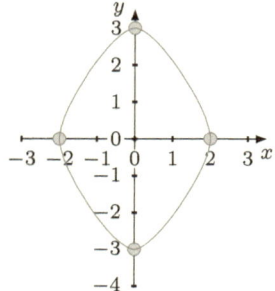

Locus : Equation to a Locus

It is clear from the sketch that the locus is an ellipse. ∎

A and B being the fixed points $(a, 0)$ and $(-a, 0)$ respectively, obtain the equations giving the locus of P, when

§ Problem 3.0.8. $PA^2 - PB^2 = a\ constant\ quantity = 2k^2.$ ◊

§§ Solution. Let (x, y) be the coordinates of the point P. Then
$$PA^2 - PB^2 = 2k^2$$
$$\therefore (x-a)^2 + (y-0)^2 - (x+a)^2 - (y-0)^2 = 2k^2$$
$$\therefore x^2 + a^2 - 2ax + y^2 - x^2 - a^2 - 2ax - y^2 = 2k^2$$
$$\therefore -4ax = 2k^2$$
$$\therefore 2ax + k^2 = 0.$$ ∎

§ Problem 3.0.9. $PA = nPB$, n being constant. ◊

§§ Solution. Let (x, y) be the coordinates of the point P. Then
$$PA = nPB$$
$$\therefore PA^2 = n^2 PB^2$$
$$\therefore (x-a)^2 + (y-0)^2 = n^2 \left\{(x+a)^2 + (y-0)^2\right\}$$
$$\therefore x^2 + a^2 - 2ax + y^2 = n^2(x^2 + a^2 + 2ax + y^2)$$
$$\therefore (n^2 - 1)(x^2 + y^2 + a^2) + 2ax(n^2 + 1) = 0.$$ ∎

§ Problem 3.0.10. $PA + PB = c$, a constant quantity. ◊

§§ Solution. Let (x, y) be the coordinates of the point P. Then
$$PA + PB = c$$
$$\therefore \sqrt{(x-a)^2 + (y-0)^2} + \sqrt{(x+a)^2 + (y-0)^2} = c$$
$$\therefore \sqrt{(x+a)^2 + y^2} + \sqrt{(x-a)^2 + y^2} = c \tag{3.1}$$

Multiplying both sides with $\left(\sqrt{(x+a)^2 + y^2} - \sqrt{(x-a)^2 + y^2}\right)$, we get

$$\therefore \left(\sqrt{(x+a)^2 + y^2} + \sqrt{(x-a)^2 + y^2}\right)$$
$$\left(\sqrt{(x+a)^2 + y^2} - \sqrt{(x-a)^2 + y^2}\right)$$
$$= c\left(\sqrt{(x+a)^2 + y^2} - \sqrt{(x-a)^2 + y^2}\right)$$
$$\therefore (x+a)^2 + y^2 - (x-a)^2 - y^2$$
$$= c\left(\sqrt{(x+a)^2 + y^2} - \sqrt{(x-a)^2 + y^2}\right)$$
$$\therefore 4ax = c\left(\sqrt{(x+a)^2 + y^2} - \sqrt{(x-a)^2 + y^2}\right)$$

$$\therefore \sqrt{(x+a)^2 + y^2} - \sqrt{(x-a)^2 + y^2} = \frac{4ax}{c} \tag{3.2}$$

Adding (3.1) and (3.2) :
$$2\sqrt{(x+a)^2 + y^2} = c + \frac{4ax}{c}$$
$$\therefore 4\left\{(x+a)^2 + y^2\right\} = \left(c + \frac{4ax}{c}\right)^2$$

Locus : Equation to a Locus

$$\therefore 4\left\{x^2 + a^2 + y^2 + 2ax\right\} = c^2 + \frac{16a^2x^2}{c^2} + 8ax$$
$$\therefore 4x^2(c^2 - 4a^2) + 4c^2y^2 = c^2(c^2 - 4a^2).$$

Alternative Solution :

$$PA + PB = c$$
$$\therefore PA = c - PB$$
$$\therefore PA^2 = (c - PB)^2 = c^2 + PB^2 - 2 \cdot c \cdot PB$$
$$\therefore 2 \cdot c \cdot PB = c^2 + PB^2 - PA^2$$
$$\therefore 2 \cdot c \cdot \sqrt{(x+a)^2 + (y-0)^2}$$
$$= c^2 + (x+a)^2 + (y-0)^2 - (x-a)^2 - (y-0)^2$$
$$\therefore 2c\sqrt{(x+a)^2 + y^2} = c^2 + 4ax$$
$$\therefore 4c^2(x^2 + a^2 + 2ax + y^2) = (c^2 + 4ax)^2 = c^4 + 8axc^2 + 16a^2x^2$$
$$\therefore 4x^2(c^2 - 4a^2) + 4c^2y^2 = c^2(c^2 - 4a^2). \qquad \blacksquare$$

§ Problem 3.0.11. $PB^2 + PC^2 = 2PA^2$, C being the point $(c, 0)$. ◊

§§ Solution. Let (x, y) be the coordinates of the point P. Then
$$PB^2 + PC^2 = 2PA^2$$
$$\therefore (x+a)^2 + (y-0)^2 + (x-c)^2 + (y-0)^2 = 2\left\{(x-a)^2 + (y-0)^2\right\}$$
$$\therefore x^2 + a^2 + 2ax + y^2 + x^2 + c^2 - 2cx + y^2 = 2(x^2 + a^2 - 2ax + y^2)$$
$$\therefore 6ax - a^2 + c^2 - 2cx = 0$$
$$\therefore (6a - 2c)x = a^2 - c^2. \qquad \blacksquare$$

§ Problem 3.0.12. *Find the locus of a point whose distance from the point $(1, 2)$ is equal to its distance from the axis of y.* ◊

§§ Solution. Let (x, y) be the coordinates of the point P. Then its distance from the axis of y is x.
$$\therefore (x-1)^2 + (y-2)^2 = x^2$$
$$\therefore x^2 + 1 - 2x + y^2 + 4 - 4y = x^2$$
$$\therefore y^2 - 4y - 2x + 5 = 0. \qquad \blacksquare$$

Find the equation to the locus of a point which is always equidistant from the points whose coordinates are

§ Problem 3.0.13. $(1, 0)$ *and* $(0, -2)$. ◊

§§ Solution. Let (x, y) be the coordinates of the point P.
$$\therefore (x-1)^2 + (y-0)^2 = (x-0)^2 + (y+2)^2$$
$$\therefore x^2 + 1 - 2x + y^2 = x^2 + y^2 + 4 + 4y$$
$$\therefore 4y + 2x + 3 = 0. \qquad \blacksquare$$

§ Problem 3.0.14. $(2, 3)$ *and* $(4, 5)$. ◊

§§ Solution. Let (x, y) be the coordinates of the point P.
$$\therefore (x-2)^2 + (y-3)^2 = (x-4)^2 + (y-5)^2$$
$$\therefore x^2 + 4 - 4x + y^2 + 9 - 6y = x^2 + 16 - 8x + y^2 + 25 - 10y$$
$$\therefore 4x + 4y = 28$$
$$\therefore x + y = 7. \qquad \blacksquare$$

§ Problem 3.0.15. $(a+b, a-b)$ *and* $(a-b, a+b)$. ◊

Locus : Equation to a Locus

§§ Solution. Let (x, y) be the coordinates of the point P.
$$\therefore (x - \overline{a+b})^2 + (y - \overline{a-b})^2 = (x - \overline{a-b})^2 + (y - \overline{a+b})^2$$
$$\therefore -2x(a+b) - 2y(a-b) = -2x(a-b) - 2y(a+b)$$
$$\therefore 2bx = 2by$$
$$\therefore y = x. \qquad \blacksquare$$

Find the equation to the locus of a point which moves so that

§ Problem 3.0.16. *its distance from the axis of x is three times its distance from the axis of y.* ◊

§§ Solution. Let (x, y) be the coordinates of the point P.
$$\therefore y = 3x. \qquad \blacksquare$$

§ Problem 3.0.17. *its distance from the point $(a, 0)$ is always four times its distance from the axis of y.* ◊

§§ Solution. Let (x, y) be the coordinates of the point P.
$$\therefore \sqrt{(x-a)^2 + (y-0)^2} = 4x$$
$$\therefore x^2 + a^2 - 2ax + y^2 = 16x^2$$
$$\therefore 15x^2 - y^2 + 2ax = a^2. \qquad \blacksquare$$

§ Problem 3.0.18. *the sum of the squares of its distances from the axes is equal to 3.* ◊

§§ Solution. Let (x, y) be the coordinates of the point P.
$$\therefore x^2 + y^2 = 3. \qquad \blacksquare$$

§ Problem 3.0.19. *the square of its distance from the point $(0, 2)$ is equal to 4.* ◊

§§ Solution. Let (x, y) be the coordinates of the point P.
$$\therefore (x - 0)^2 + (y - 2)^2 = 4$$
$$\therefore x^2 + y^2 + 4 - 4y = 4$$
$$\therefore x^2 + y^2 = 4y. \qquad \blacksquare$$

§ Problem 3.0.20. *its distance from the point $(3, 0)$ is three times its distance from $(0, 2)$.* ◊

§§ Solution. Let (x, y) be the coordinates of the point P.
$$\therefore \sqrt{(x-3)^2 + (y-0)^2} = 3\sqrt{(x-0)^2 + (y-2)^2}$$
$$\therefore x^2 + 9 - 6x + y^2 = 9(x^2 + y^2 + 4 - 4y)$$
$$\therefore 8x^2 + 8y^2 + 6x - 36y + 27 = 0. \qquad \blacksquare$$

§ Problem 3.0.21. *its distance from the axis of x is always one half its distance from the origin.* ◊

§§ Solution. Let (x, y) be the coordinates of the point P.
$$\therefore y = \frac{1}{2}\sqrt{(x-0)^2 + (y-0)^2}$$
$$\therefore 4y^2 = x^2 + y^2$$
$$\therefore x^2 = 3y^2. \qquad \blacksquare$$

Locus : Equation to a Locus

§ Problem 3.0.22. *A fixed point is at a perpendicular distance a from a fixed straight line and a point moves so that its distance from the fixed point is always equal to its distance from the fixed line. Find the equation to its locus, the axes of coordinates being drawn through the fixed point and being parallel and perpendicular to the given line.* ◊

§§ Solution. Let $O(0,0)$ be the fixed point, MN be the fixed straight line and $P(x,y)$ be the moving point.

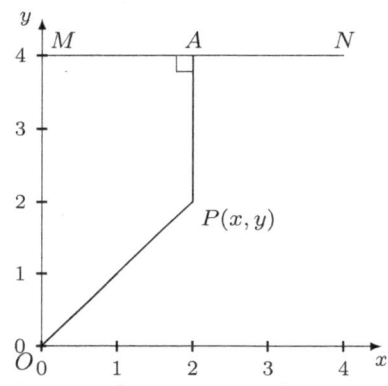

It is given that $OM = a$ and $OP = PA$.

$$\because OP = PA$$
$$\therefore \sqrt{(x-0)^2 + (y-0)^2} = a - y$$
$$\therefore x^2 + y^2 = (a-y)^2 = a^2 + y^2 - 2ay$$
$$\therefore x^2 + 2ay = a^2.$$

∎

§ Problem 3.0.23. *In the previous question if the first distance be (1), always half, and (2), always twice, the second distance, find the equations to the respective loci.* ◊

§§ Solution. (1)

$$\because OP = \frac{1}{2}PA$$
$$\therefore \sqrt{x^2 + y^2} = \frac{1}{2}(a-y)$$
$$\therefore x^2 + y^2 = \frac{1}{4}(a-y)^2$$
$$\therefore 4x^2 + 4y^2 = a^2 + y^2 - 2ay$$
$$\therefore 4x^2 + 3y^2 + 2ay = a^2.$$

(2)

$$\because OP = 2PA$$
$$\therefore \sqrt{x^2 + y^2} = 2(a-y)$$
$$\therefore x^2 + y^2 = 4(a-y)^2 = 4a^2 + 4y^2 - 8ay$$
$$\therefore x^2 - 3y^2 + 8ay = 4a^2.$$

∎

Chapter 4

The Straight Line : Rectangular Coordinates

4.1 Equations, Slope and Intercepts

Find the equation to the straight line

§ Problem 4.1.1. *cutting off an intercept unity from the positive direction of the axis of y and inclined at* $45°$ *to the axis of x.* ◊

§§ Solution. It is given that $c = 1$ and $\alpha = 45°$.
$$\therefore m = \tan\alpha = \tan 45° = 1.$$
$\therefore$ the equation to the straight line is $y = mx + c$
$$\therefore y = 1 \cdot x + 1$$
$$\therefore y = x + 1.$$
∎

§ Problem 4.1.2. *cutting off an intercept* -5 *from the axis of y and being equally inclined to the axes.* ◊

§§ Solution. It is given that $c = -5$ and $\alpha = 45°$.
$$\therefore m = \tan\alpha = \tan 45° = 1.$$
$\therefore$ the equation to the straight line is $y = mx + c$
$$\therefore y = 1 \cdot x + (-5)$$
$$\therefore y = x - 5$$
$$\therefore x - y - 5 = 0.$$
∎

§ Problem 4.1.3. *cutting off an intercept* 2 *from the negative direction of the axis of y and inclined at* $30°$ *to* OX. ◊

§§ Solution. It is given that $c = -2$ and $\alpha = 30°$.
$$\therefore m = \tan\alpha = \tan 30° = \frac{1}{\sqrt{3}}.$$

4.1. Equations, Slope and Intercepts

∴ the equation to the straight line is $y = mx + c$

$$\therefore y = \frac{1}{\sqrt{3}} \cdot x + (-2)$$

$$\therefore y = \frac{x}{\sqrt{3}} - 2$$

$$\therefore x - y\sqrt{3} - 2\sqrt{3} = 0.$$ ∎

§ Problem 4.1.4. *cutting off an intercept -3 from the axis of y and inclined at an angle $\tan^{-1}\frac{3}{5}$ to the axis of x.* ◊

§§ Solution. It is given that $c = -3$ and $\alpha = \tan^{-1}\frac{3}{5}$.

$$\therefore m = \tan\alpha = \frac{3}{5}.$$

∴ the equation to the straight line is $y = mx + c$

$$\therefore y = \frac{3}{5} \cdot x + (-3)$$

$$\therefore 5y - 3x + 15 = 0.$$ ∎

Find the equation to the straight line

§ Problem 4.1.5. *cutting off intercepts 3 and 2 from the axes.* ◊

§§ Solution. It is given that $a = 3$ and $b = 2$.

∴ the equation to the straight line is

$$\frac{x}{a} + \frac{y}{b} = 1$$

$$\therefore \frac{x}{3} + \frac{y}{2} = 1$$

$$\therefore 2x + 3y = 6.$$ ∎

§ Problem 4.1.6. *cutting off intercepts -5 and 6 from the axes.* ◊

§§ Solution. It is given that $a = -5$ and $b = 6$.

∴ the equation to the straight line is

$$\frac{x}{a} + \frac{y}{b} = 1$$

$$\therefore \frac{x}{-5} + \frac{y}{6} = 1$$

$$\therefore 6x - 5y + 30 = 0.$$ ∎

§ Problem 4.1.7. *Find the equation to the straight line which passes through the point $(5, 6)$ and has intercepts on the axes*

1. *equal in magnitude and both positive,*

2. *equal in magnitude but opposite in sign.* ◊

§§ Solution. 1. Let us assume that the length of the intercept is a. Then the equation of the straight line is

$$\frac{x}{a} + \frac{y}{a} = 1$$

$$\therefore x + y = a.$$

Since it passes through the point $(5, 6)$,

$$\therefore 5 + 6 = a$$

$$\therefore a = 11.$$

Hence the equation becomes $x + y = 11$.

4.1. Equations, Slope and Intercepts

2. Let us assume that the intercept on the x-axis is a. and the intercept on the y-axis is $-a$. Then the equation of the straight line is
$$\frac{x}{a} + \frac{y}{-a} = 1$$
$$\therefore x - y = a.$$

Since it passes through the point $(5, 6)$,
$$\therefore 5 - 6 = a$$
$$\therefore a = -1.$$

Hence the equation becomes $y - x = 1$. ∎

§ Problem 4.1.8. *Find the equations to the straight lines which pass through the point $(1, -2)$ and cut off equal distances from the two axes.* ◊

§§ Solution. It is not difficult to see that the equations to the straight lines are
$$\frac{x}{a} \pm \frac{y}{a} = 1$$
$$\therefore x \pm y = a.$$

And since these pass through the point $(1, -2)$, hence
$$1 \mp 2 = a$$
$$\therefore a = -1 \text{ or } 3.$$

Hence the equations are
$$x + y + 1 = 0 \text{ and}$$
$$x - y = 3.$$ ∎

§ Problem 4.1.9. *Find the equation to the straight line which passes through the given point (x', y') and is such that the given point bisects the part intercepted between the axes.* ◊

§§ Solution. As can be seen that the part of the straight line intercepted between the axes is AB, with the coordinates as $A(a, 0)$ and $B(0, b)$ so that the intercept on x-axis is a and the intercept on y-axis is b.

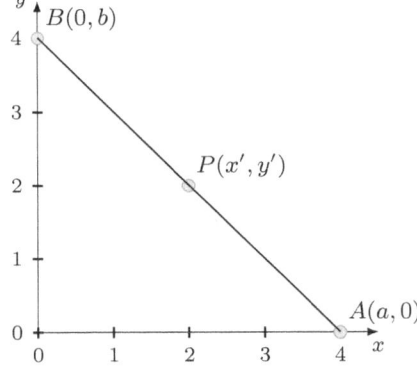

4.1. Equations, Slope and Intercepts

Hence the equation to the straight line is
$$\frac{x}{a} + \frac{y}{b} = 1 \tag{4.1}$$

It is given that the given point $P(x', y')$ bisects the line AB. Hence $P(x', y')$ is the mid-point of $A(a, 0)$ and $B(0, b)$.

$$\therefore x' = \frac{a+0}{2} = \frac{a}{2}$$
$$\therefore a = 2x'$$
$$\therefore y' = \frac{0+b}{2} = \frac{b}{2}$$
$$\therefore b = 2y'$$

Putting these values of a and b into the equation (4.1), we get the equation to the straight line as :
$$\frac{x}{2x'} + \frac{y}{2y'} = 1$$
$$\therefore xy' + yx' = 2x'y'. \qquad \blacksquare$$

§ Problem 4.1.10. *Find the equation to the straight line which passes through the point $(-4, 3)$ and is such that the portion of it between the axes is divided by the point in the ratio 5 : 3.* ◊

§§ Solution. It is given that the given point $P(-4, 3)$ divides the line AB in the ratio 5 : 3, where the coordinates of A and B are $A(a, 0)$ and $B(0, b)$.

$$\therefore -4 = \frac{5 \cdot 0 + 3 \cdot a}{5 + 3}$$
$$\therefore a = -\frac{32}{3}.$$
$$\therefore 3 = \frac{5 \cdot b + 3 \cdot 0}{5 + 3}$$
$$\therefore b = \frac{24}{5}.$$

Therefore the equation to the straight line is
$$\frac{x}{a} + \frac{y}{b} = 1$$
$$\therefore \frac{x}{-\frac{32}{3}} + \frac{y}{\frac{24}{5}} = 1$$
$$\therefore 20y - 9x = 96. \qquad \blacksquare$$

Trace the straight lines whose equations are

§ Problem 4.1.11. $x + 2y + 3 = 0$. ◊

§§ Solution. The equation to the straight line is
$$x + 2y + 3 = 0.$$

Putting $x = 0$, we have $y = -\frac{3}{2} = -1.5$, so the point $P(0, -1.5)$ is on this line.

Similarly, putting $y = 0$, we have $x = -3$, so the point $Q(-3, 0)$ is on this line.

4.1. Equations, Slope and Intercepts 47

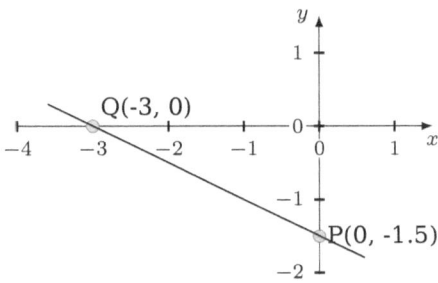

§ **Problem 4.1.12.** 5x - 7y - 9 = 0. ◊
§§ **Solution**. The equation to the straight line is
$$5x - 7y - 9 = 0$$
$$\therefore y = \frac{5x - 9}{7}$$

Putting $x = -1$, we have $y = -\frac{14}{7} = -2$, so the point $P(-1, -2)$ is on this line.

Similarly, putting $x = 6$, we have $y = 3$, so the point $Q(6, 3)$ is on this line.

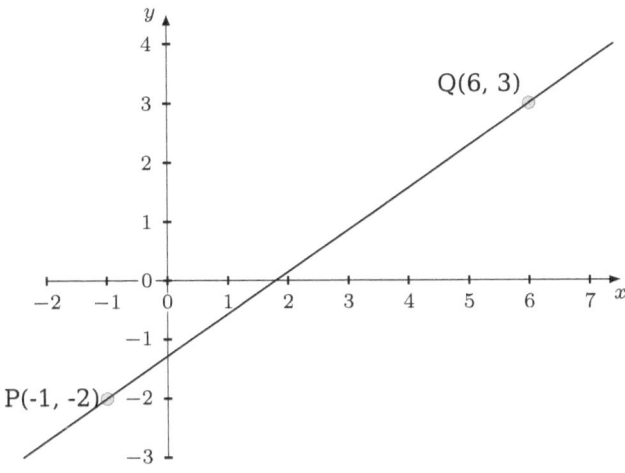

§ **Problem 4.1.13.** 3x + 7y = 0. ◊
§§ **Solution**. The equation to the straight line is
$$3x + 7y = 0$$
$$\therefore y = -\frac{3x}{7}$$

Putting $x = 0$, we have $y = 0$, so the point $P(0, 0)$ is on this line.
Similarly, putting $x = -7$, we have $y = 3$, so the point $Q(-7, 3)$ is on this line.

4.1. Equations, Slope and Intercepts   48

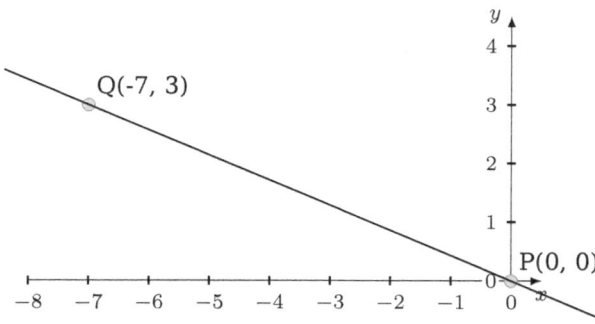

§ Problem 4.1.14. $3x + 7y = 0$. ◊

§§ Solution. The equation to the straight line is
$$2x - 3y + 4 = 0$$
$$\therefore y = \frac{2x + 4}{3}$$

Putting $x = -2$, we have $y = 0$, so the point $P(-2, 0)$ is on this line.

Similarly, putting $x = 1$, we have $y = 2$, so the point $Q(1, 2)$ is on this line.

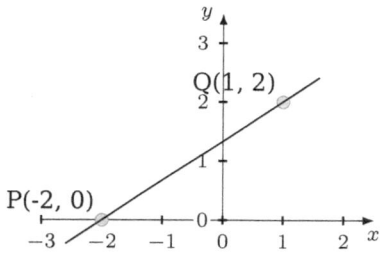

Find the equations to the straight lines passing through the following pairs of points.

§ Problem 4.1.15. $(0, 0)$ and $(2, -2)$. ◊

§§ Solution. The equation to the straight line is :
$$y - 0 = \frac{-2 - 0}{2 - 0}(x - 0)$$
$$\therefore y = -x$$
$$\therefore x + y = 0.$$ ∎

§ Problem 4.1.16. $(3, 4)$ and $(5, -6)$. ◊

§§ Solution. The equation to the straight line is :
$$y - 4 = \frac{6 - 4}{5 - 3}(x - 3)$$
$$\therefore y - 4 = x - 3$$
$$\therefore y - x = 1.$$ ∎

§ Problem 4.1.17. $(-1, 3)$ and $(6, -7)$. ◊

4.1. Equations, Slope and Intercepts

§§ Solution. The equation to the straight line is :
$$y - 3 = \frac{-7 - 3}{6 - (-1)}(x - (-1))$$
$$\therefore y - 3 = -\frac{10}{7}(x + 1)$$
$$\therefore 7y + 10x = 11.$$ ∎

§ Problem 4.1.18. $(0, -a)$ and $(b, 0)$.   ◊
§§ Solution. The equation to the straight line is :
$$y - (-a) = \frac{0 - (-a)}{b - 0}(x - 0)$$
$$\therefore y + a = \frac{a}{b}x$$
$$\therefore ax - by = ab.$$ ∎

§ Problem 4.1.19. (a, b) and $(a+b, a-b)$.   ◊
§§ Solution. The equation to the straight line is :
$$y - b = \frac{(a-b) - b}{(a+b) - a}(x - a)$$
$$\therefore y - b = \frac{a - 2b}{b}(x - a)$$
$$\therefore by - b^2 = (a - 2b)x + 2ab - a^2$$
$$\therefore (a - 2b)x - by + b^2 + 2ab - a^2 = 0.$$ ∎

§ Problem 4.1.20. $\left(at_1^2, 2at_1\right)$ and $\left(at_2^2, 2at_2\right)$.   ◊
§§ Solution. The equation to the straight line is :
$$y - 2at_1 = \frac{2at_2 - 2at_1}{at_2^2 - at_1^2}(x - at_1^2)$$
$$\therefore y - 2at_1 = \frac{2(t_2 - t_1)}{(t_2 + t_1)(t_2 - t_1)}(x - at_1^2)$$
$$\therefore (y - 2at_1)(t_1 + t_2) = 2x - 2at_1^2$$
$$\therefore y(t_1 + t_2) - 2x = 2at_1 t_2.$$ ∎

§ Problem 4.1.21. $\left(at_1, \dfrac{a}{t_1}\right)$ and $\left(at_2, \dfrac{a}{t_2}\right)$.   ◊
§§ Solution. The equation to the straight line is :
$$y - \frac{a}{t_1} = \frac{\frac{a}{t_2} - \frac{a}{t_1}}{at_2 - at_1}(x - at_1)$$
$$\therefore y - \frac{a}{t_1} = \frac{(t_1 - t_2)}{(t_2 - t_1)t_1 t_2}(x - at_1)$$
$$\therefore t_1 t_2 y - at_2 = at_1 - x$$
$$\therefore t_1 t_2 y + x = a(t_1 + t_2).$$ ∎

§ Problem 4.1.22. $(a\cos\phi_1, a\sin\phi_1)$ and $(a\cos\phi_2, a\sin\phi_2)$.   ◊

4.1. Equations, Slope and Intercepts

§§ Solution. The equation to the straight line is :
$$y - a\sin\phi_1 = \frac{a\sin\phi_2 - a\sin\phi_1}{a\cos\phi_2 - a\cos\phi_1}(x - a\cos\phi_1)$$

$$\therefore y - a\sin\phi_1 = \frac{2\cos\dfrac{\phi_2+\phi_1}{2}\sin\dfrac{\phi_2-\phi_1}{2}}{-2\sin\dfrac{\phi_2+\phi_1}{2}\sin\dfrac{\phi_2-\phi_1}{2}}(x - a\cos\phi_1)$$

$$\therefore y\sin\frac{\phi_1+\phi_2}{2} - a\sin\phi_1\sin\frac{\phi_1+\phi_2}{2} = a\cos\phi_1\cos\frac{\phi_1+\phi_2}{2} - x\cos\frac{\phi_1+\phi_2}{2}$$

$$\therefore x\cos\frac{\phi_1+\phi_2}{2} + y\sin\frac{\phi_1+\phi_2}{2} = a\cos\left(\phi_1 - \frac{\phi_1+\phi_2}{2}\right)$$

$$\therefore x\cos\frac{\phi_1+\phi_2}{2} + y\sin\frac{\phi_1+\phi_2}{2} = a\cos\frac{\phi_1-\phi_2}{2}.$$

Alternative Solution :

Let the required equation be
$$y = mx + c \tag{4.2}$$

Since (4.2) passes through the first point $(a\cos\phi_1, a\sin\phi_1)$, we have
$$a\sin\phi_1 = ma\cos\phi_1 + c$$
$$\therefore c = a\sin\phi_1 - ma\cos\phi_1$$

Hence (4.2) becomes
$$y = mx + a\sin\phi_1 - ma\cos\phi_1 \tag{4.3}$$

Since this line also passes through the second point
$$(a\cos\phi_2, a\sin\phi_2),$$

we have
$$a\sin\phi_2 = ma\cos\phi_2 + a\sin\phi_1 - ma\cos\phi_1$$

$$\therefore m = \frac{\sin\phi_1 - \sin\phi_2}{\cos\phi_1 - \cos\phi_2}$$

$$\therefore m = \frac{2\cos\dfrac{\phi_1+\phi_2}{2}\sin\dfrac{\phi_1-\phi_2}{2}}{-2\sin\dfrac{\phi_1+\phi_2}{2}\sin\dfrac{\phi_1-\phi_2}{2}}$$

$$\therefore m = -\frac{\cos\dfrac{\phi_1+\phi_2}{2}}{\sin\dfrac{\phi_1+\phi_2}{2}}$$

Hence (4.3) becomes
$$y = -\frac{\cos\dfrac{\phi_1+\phi_2}{2}}{\sin\dfrac{\phi_1+\phi_2}{2}}x + a\sin\phi_1 + \frac{\cos\dfrac{\phi_1+\phi_2}{2}}{\sin\dfrac{\phi_1+\phi_2}{2}}\cos\phi_1$$

$$\therefore x\cos\frac{\phi_1+\phi_2}{2} + y\sin\frac{\phi_1+\phi_2}{2} = a\cos\phi_1\cos\frac{\phi_1+\phi_2}{2} + a\sin\phi_1\sin\frac{\phi_1+\phi_2}{2}$$

$$= a\cos\left(\phi_1 - \frac{\phi_1+\phi_2}{2}\right) = a\cos\frac{\phi_1-\phi_2}{2}$$

$$\therefore x\cos\frac{\phi_1+\phi_2}{2} + y\sin\frac{\phi_1+\phi_2}{2} = a\cos\frac{\phi_1-\phi_2}{2}.$$

4.1. Equations, Slope and Intercepts 51

Alternative Solution :
Let the required equation to the straight line be
$$lx + my = 1 \tag{4.4}$$
Since this passes through the given points, hence we have
$$la\cos\phi_1 + ma\sin\phi_1 = 1 \tag{4.5}$$
$$la\cos\phi_2 + ma\sin\phi_2 = 1 \tag{4.6}$$
$(4.6) \times \sin\phi_1 - (4.5) \times \sin\phi_2 \implies$
$$la\sin\phi_1\cos\phi_2 - la\cos\phi_1\sin\phi_2 = \sin\phi_1 - \sin\phi_2$$
$$\therefore la\sin(\phi_1 - \phi_2) = \sin\phi_1 - \sin\phi_2$$
$$\therefore 2la\sin\frac{\phi_1-\phi_2}{2}\cos\frac{\phi_1-\phi_2}{2} = 2\cos\frac{\phi_1+\phi_2}{2}\sin\frac{\phi_1-\phi_2}{2}$$
$$\therefore l = \frac{\cos\dfrac{\phi_1+\phi_2}{2}}{a\cos\dfrac{\phi_1-\phi_2}{2}} \tag{4.7}$$
$(4.5) \times \cos\phi_2 - (4.6) \times \cos\phi_1 \implies$
$$ma(\sin\phi_1\cos\phi_2 - \cos\phi_1\sin\phi_2) = \cos\phi_2 - \cos\phi_1$$
$$\therefore mb\sin(\phi_1 - \phi_2) = \cos\phi_2 - \cos\phi_1$$
$$2ma\sin\frac{\phi_1-\phi_2}{2}\cos\frac{\phi_1-\phi_2}{2} = 2\sin\frac{\phi_1+\phi_2}{2}\sin\frac{\phi_1-\phi_2}{2}$$
$$\therefore m = \frac{\sin\dfrac{\phi_1+\phi_2}{2}}{a\cos\dfrac{\phi_1-\phi_2}{2}} \tag{4.8}$$
Putting the values of l and m in (4.4), the equation becomes :
$$x\cos\frac{\phi_1+\phi_2}{2} + y\sin\frac{\phi_1+\phi_2}{2} = a\cos\frac{\phi_1-\phi_2}{2}. \qquad\blacksquare$$

§ Problem 4.1.23. $(a\cos\phi_1, b\sin\phi_1)$ *and* $(a\cos\phi_2, b\sin\phi_2)$. $\diamond$
§§ Solution. Let the required equation to the straight line be
$$lx + my = 1 \tag{4.9}$$
Since this passes through the given points, hence we have
$$la\cos\phi_1 + mb\sin\phi_1 = 1 \tag{4.10}$$
$$la\cos\phi_2 + mb\sin\phi_2 = 1 \tag{4.11}$$
$(4.11) \times \sin\phi_1 - (4.10) \times \sin\phi_2 \implies$
$$la\sin\phi_1\cos\phi_2 - la\cos\phi_1\sin\phi_2 = \sin\phi_1 - \sin\phi_2$$
$$\therefore la\sin(\phi_1 - \phi_2) = \sin\phi_1 - \sin\phi_2$$
$$\therefore 2la\sin\frac{\phi_1-\phi_2}{2}\cos\frac{\phi_1-\phi_2}{2} = 2\cos\frac{\phi_1+\phi_2}{2}\sin\frac{\phi_1-\phi_2}{2}$$
$$\therefore l = \frac{\cos\dfrac{\phi_1+\phi_2}{2}}{a\cos\dfrac{\phi_1-\phi_2}{2}} \tag{4.12}$$
$(4.10) \times \cos\phi_2 - (4.11) \times \cos\phi_1 \implies$
$$mb(\sin\phi_1\cos\phi_2 - \cos\phi_1\sin\phi_2) = \cos\phi_2 - \cos\phi_1$$
$$\therefore mb\sin(\phi_1 - \phi_2) = \cos\phi_2 - \cos\phi_1$$
$$2mb\sin\frac{\phi_1-\phi_2}{2}\cos\frac{\phi_1-\phi_2}{2} = 2\sin\frac{\phi_1+\phi_2}{2}\sin\frac{\phi_1-\phi_2}{2}$$

4.1. Equations, Slope and Intercepts

$$\therefore m = \frac{\sin \frac{\phi_1 + \phi_2}{2}}{b \cos \frac{\phi_1 - \phi_2}{2}} \qquad (4.13)$$

Putting the values of l and m in (4.9), the equation becomes :

$$\frac{x}{a} \cos \frac{\phi_1 + \phi_2}{2} + \frac{y}{b} \sin \frac{\phi_1 + \phi_2}{2} = \cos \frac{\phi_1 - \phi_2}{2}.$$

Alternative Solution :

The equation to the straight line is

$$y - b \sin \phi_1 = \frac{b \sin \phi_2 - b \sin \phi_1}{a \cos \phi_2 - a \cos \phi_1}(x - a \cos \phi_1)$$

$$\therefore y - b \sin \phi_1 = \frac{2b \cos \frac{\phi_2 + \phi_1}{2} \sin \frac{\phi_2 - \phi_1}{2}}{-2a \sin \frac{\phi_2 + \phi_1}{2} \sin \frac{\phi_2 - \phi_1}{2}}(x - a \cos \phi_1)$$

$$\therefore ay \sin \frac{\phi_1 + \phi_2}{2} - ab \sin \phi_1 \sin \frac{\phi_2 + \phi_1}{2}$$

$$= -bx \cos \frac{\phi_2 + \phi_1}{2} + ab \cos \phi_1 \cos \frac{\phi_2 + \phi_1}{2}$$

$$\therefore xb \cos \frac{\phi_2 + \phi_1}{2} + yb \sin \frac{\phi_2 + \phi_1}{2} = ab \cos \left(\phi_1 - \frac{\phi_2 + \phi_1}{2} \right)$$

$$\therefore \frac{x}{a} \cos \frac{\phi_1 + \phi_2}{2} + \frac{y}{b} \sin \frac{\phi_1 + \phi_2}{2} = \cos \frac{\phi_1 - \phi_2}{2}. \qquad \blacksquare$$

§ Problem 4.1.24. $(a \sec \phi_1, b \tan \phi_1)$ *and* $(a \sec \phi_2, b \tan \phi_2)$. $\diamond$

§§ Solution. The required equation to the straight line is

$$y - b \tan \phi_1 = \frac{b \tan \phi_2 - b \tan \phi_1}{a \sec \phi_2 - a \sec \phi_1}(x - a \sec \phi_1)$$

$$\therefore \frac{y - b \tan \phi_1}{x - a \sec \phi_1} = \frac{b \left(\dfrac{\sin \phi_2}{\cos \phi_2} - \dfrac{\sin \phi_1}{\cos \phi_1} \right)}{a \left(\dfrac{1}{\cos \phi_2} - \dfrac{1}{\cos \phi_1} \right)}$$

$$\therefore \frac{y - b \tan \phi_1}{x - a \sec \phi_1} = \frac{b(\sin \phi_2 \cos \phi_1 - \sin \phi_1 \cos \phi_2)}{a(\cos \phi_1 - \cos \phi_2)}$$

$$= \frac{b \sin(\phi_2 - \phi_1)}{a(\cos \phi_1 - \cos \phi_2)}$$

$$= \frac{2b \sin \dfrac{\phi_2 - \phi_1}{2} \cos \dfrac{\phi_2 - \phi_1}{2}}{2a \sin \dfrac{\phi_1 + \phi_2}{2} \sin \dfrac{\phi_2 - \phi_1}{2}} = \frac{b \cos \dfrac{\phi_1 - \phi_2}{2}}{a \sin \dfrac{\phi_1 + \phi_2}{2}}$$

$$\therefore bx \cos \left(\frac{\phi_1 - \phi_2}{2} \right) - ay \sin \left(\frac{\phi_1 + \phi_2}{2} \right)$$

$$= ab \sec \phi_1 \left\{ \cos \frac{\phi_1 - \phi_2}{2} - \sin \phi_1 \sin \frac{\phi_1 + \phi_2}{2} \right\} \qquad (4.14)$$

$$\because \sin a \sin b = \frac{1}{2}[\cos(a - b) - \cos(a + b)]$$

$$\therefore \sin \phi_1 \sin \frac{\phi_1 + \phi_2}{2} = \frac{1}{2} \left[\cos \left(\phi_1 - \frac{\phi_1 + \phi_2}{2} \right) - \cos \left(\phi_1 + \frac{\phi_1 + \phi_2}{2} \right) \right]$$

4.1. Equations, Slope and Intercepts

$$= \frac{1}{2}\left[\cos\left(\frac{\phi_1 - \phi_2}{2}\right) - \cos\left(\frac{3\phi_1 + \phi_2}{2}\right)\right]$$

With this, (4.14) becomes
$$bx\cos\left(\frac{\phi_1 - \phi_2}{2}\right) - ay\sin\left(\frac{\phi_1 + \phi_2}{2}\right)$$
$$= \frac{ab}{2}\sec\phi_1\left\{\cos\frac{\phi_1 - \phi_2}{2} + \cos\frac{3\phi_1 + \phi_2}{2}\right\} \quad (4.15)$$

$$\because \cos a + \cos b = 2\cos\frac{a+b}{2}\cos\frac{a-b}{2}$$
$$\therefore \cos\frac{\phi_1 - \phi_2}{2} + \cos\frac{3\phi_1 + \phi_2}{2}$$
$$= 2\left(\cos\frac{\frac{\phi_1-\phi_2}{2} + \frac{3\phi_1+\phi_2}{2}}{2}\right)\cos\left(\frac{\frac{\phi_1-\phi_2}{2} - \frac{3\phi_1+\phi_2}{2}}{2}\right)$$
$$\therefore \cos\frac{\phi_1 - \phi_2}{2} + \cos\frac{3\phi_1 + \phi_2}{2} = 2\cos\phi_1\cos\frac{\phi_1 + \phi_2}{2}$$

With this, (4.15) becomes
$$bx\cos\left(\frac{\phi_1 - \phi_2}{2}\right) - ay\sin\left(\frac{\phi_1 + \phi_2}{2}\right) = ab\cos\frac{\phi_1 + \phi_2}{2}. \quad\blacksquare$$

Find the equations to the sides of the triangles the coordinates of whose angular points are respectively

§ Problem 4.1.25. $(1, 4)$, $(2, -3)$ and $(-1, -2)$. ◊

§§ Solution. Let us denote the angular points as $A(1, 4)$, $B(2, -3)$ and $C(-1, -2)$ respectively.

Then the equation to the side AB is
$$y - 4 = \frac{-3 - 4}{2 - 1}(x - 1) = -7x + 7$$
$$\therefore y + 7x = 11.$$

Then the equation to the side BC is
$$y - (-3) = \frac{-2 - (-3)}{-1 - 2}(x - 2) = -\frac{1}{3}(x - 2)$$
$$\therefore 3y + 9 = 2 - x$$
$$\therefore x + 3y + 7 = 0.$$

Then the equation to the side CA is
$$y - (-2) = \frac{4 - (-2)}{1 - (-1)}(x - (-1)) = 3x + 3$$
$$\therefore y - 3x = 1. \quad\blacksquare$$

§ Problem 4.1.26. $(0, 1)$, $(2, 0)$ and $(-1, -2)$. ◊

§§ Solution. Let us denote the angular points as $A(0, 1)$, $B(2, 0)$ and $C(-1, -2)$ respectively.

Then the equation to the side AB is
$$y - 1 = \frac{0 - 1}{2 - 0}(x - 0) = -\frac{x}{2}$$
$$\therefore x + 2y = 2.$$

Then the equation to the side BC is
$$y - 0 = \frac{-2 - 0}{-1 - 2}(x - 2) = \frac{2}{3}(x - 2)$$
$$\therefore 2x - 3y = 4.$$

4.1. Equations, Slope and Intercepts

Then the equation to the side CA is
$$y - (-2) = \frac{1-(-2)}{0-(-1)}(x-(-1)) = 3x+3$$
$$\therefore y - 3x = 1.$$ ∎

§ Problem 4.1.27. *Find the equations to the diagonals of the rectangle the equations of whose sides are $x = a$, $x = a'$, $y = b$, and $y = b'$.* ◊

§§ Solution. Let us draw the sides of the rectangle:

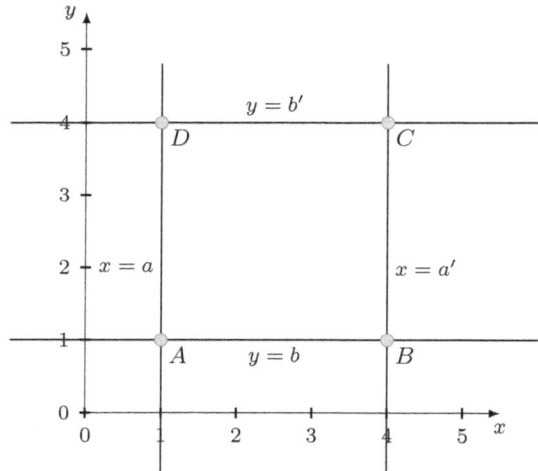

It is easy to see that the coordinates of the four corners of the rectangle $ABCD$ are
$$A(a,b); B(a',b); C(a',b'); D(a,b')$$
The equation of the diagonal AC is
$$y - b = \frac{b'-b}{a'-a}(x-a)$$
$$\therefore y(a'-a) - x(b'-b) = a'b - ab'.$$
The equation of the diagonal BD is
$$y - b = \frac{b'-b}{a-a'}(x-a')$$
$$\therefore y(a-a') - x(b'-b) = ab - a'b'$$
$$\therefore y(a'-a) + x(b'-b) = a'b' - ab.$$ ∎

§ Problem 4.1.28. *Find the equation to the straight line which bisects the distance between the points (a,b) and (a',b') and also bisects the distance between the points $(-a,b)$ and $(a',-b')$.* ◊

§§ Solution. It is given that the line passes through the points
$$A\left(\frac{a+a'}{2}, \frac{b+b'}{2}\right)$$
and
$$B\left(\frac{a'-a}{2}, \frac{b-b'}{2}\right).$$

4.1. Equations, Slope and Intercepts

Hence the equation to the straight AB is

$$y - \frac{b+b'}{2} = \frac{\frac{b-b'}{2} - \frac{b+b'}{2}}{\frac{a'-a}{2} - \frac{a+a'}{2}}\left(x - \frac{a+a'}{2}\right) = \frac{b'}{a}\left(x - \frac{a+a'}{2}\right)$$

$$\therefore 2ay - ab - ab' = 2b'x - ab' - a'b'$$
$$\therefore 2ay - 2b'x = ab - a'b'. \qquad \blacksquare$$

§ Problem 4.1.29. *Find the equations to the straight lines which go through the origin and trisect the portion of the straight line $3x + y = 12$ which is intercepted between the axes of coordinates.* ◊

§§ Solution. The equation to the given straight line is:

$$3x + y = 12$$
$$\therefore \frac{x}{4} + \frac{y}{12} = 1.$$

It's x-intercept is 4 and y-intercept is 12, i.e., it meets the x-axis at $A(4,0)$ and the y-axis at $B(0,12)$.

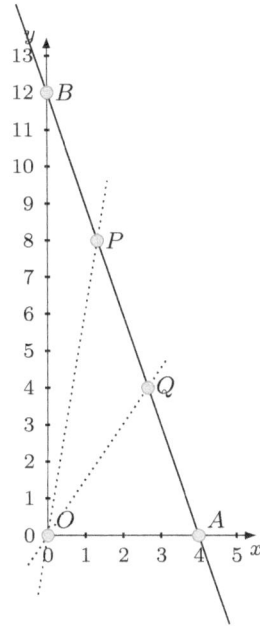

The coordinates of the points P and Q which trisects the line AB are

$$P\left(\frac{2\cdot 0 + 1\cdot 4}{2+1}, \frac{2\cdot 12 + 1\cdot 0}{2+1}\right) \equiv P\left(\frac{4}{3}, 8\right)$$

and

$$Q\left(\frac{1\cdot 0 + 2\cdot 4}{1+2}, \frac{1\cdot 12 + 2\cdot 0}{1+2}\right) \equiv Q\left(\frac{8}{3}, 4\right)$$

4.2. Angles between straight lines

Hence the equation to the straight line passing through the origin, i.e. $O(0,0)$ and $P\left(\frac{4}{3},8\right)$ is

$$y - 0 = \frac{8-0}{\frac{4}{3}-0}(x-0)$$
$$\therefore y = 6x.$$

and the equation to the straight line passing through the origin, i.e. $O(0,0)$ and $Q\left(\frac{8}{3},4\right)$ is

$$y - 0 = \frac{4-0}{\frac{8}{3}-0}(x-0)$$
$$\therefore 2y = 3x. \qquad \blacksquare$$

4.2 Angles between straight lines

Find the angles between the pairs of straight lines

§ Problem 4.2.1. $x - y\sqrt{3} = 5$ and $x\sqrt{3} + y = 7$. ◊

§§ Solution. The angle is $= \tan^{-1} \frac{m_1 - m_2}{1 + m_1 m_2}$.

The given equations are

$$x - y\sqrt{3} = 5, \therefore y = \frac{1}{\sqrt{3}}x - \frac{5}{\sqrt{3}}, \therefore m_1 = \frac{1}{\sqrt{3}}$$
$$\sqrt{3}x + y = 7, \therefore y = -\sqrt{3}x + \frac{7}{\sqrt{3}}, \therefore m_2 = -\sqrt{3}$$

Hence the angle between these two lines is

$$= \tan^{-1} \frac{m_1 - m_2}{1 + m_1 m_2} = \tan^{-1} \frac{\frac{1}{\sqrt{3}} + \sqrt{3}}{1 - \frac{1}{\sqrt{3}} \cdot \sqrt{3}} = \tan^{-1} \infty = 90°. \qquad \blacksquare$$

§ Problem 4.2.2. $x - 4y = 3$ and $6x - y = 11$. ◊

§§ Solution. The given equations are

$$x - 4y = 3, \therefore y = \frac{1}{4}x - \frac{3}{4}, \therefore m_1 = \frac{1}{4}$$
$$6x - y = 11, \therefore y = 6x - 11, \therefore m_2 = 6$$

Hence the angle between these two lines is

$$= \tan^{-1} \frac{m_1 - m_2}{1 + m_1 m_2} = \tan^{-1} \frac{\frac{1}{4} - 6}{1 + \frac{1}{4} \cdot 6} = \tan^{-1} \frac{23}{10}. \qquad \blacksquare$$

§ Problem 4.2.3. $x - 4y = 3$ and $6x - y = 11$. ◊

§§ Solution. The given equations are

$$y = 3x + 7, \therefore m_1 = 3$$
$$3y - x = 8, \therefore y = \frac{1}{3}x + \frac{8}{3}, \therefore m_2 = \frac{1}{3}$$

4.2. Angles between straight lines

Hence the angle between these two lines is

$$= \tan^{-1} \frac{m_1 - m_2}{1 + m_1 m_2} = \tan^{-1} \frac{3 - \frac{1}{3}}{1 + 3 \cdot \frac{1}{3}} = \tan^{-1} \frac{4}{3}.$$ ∎

§ Problem 4.2.4. $y = (2 - \sqrt{3})x + 5$ and $y = (2 + \sqrt{3})x - 7$. ◇

§§ Solution. The given equations are
$$y = (2 - \sqrt{3})x + 5, \therefore m_1 = 2 - \sqrt{3}$$
$$y = (2 + \sqrt{3})x - 7, \therefore m_2 = 2 + \sqrt{3}$$
Hence the angle between these two lines is
$$= \tan^{-1} \frac{m_1 - m_2}{1 + m_1 m_2} = \tan^{-1} \frac{(2 - \sqrt{3}) - (2 + \sqrt{3})}{1 + (2 - \sqrt{3})(2 + \sqrt{3})}$$
$$= \tan^{-1} \frac{-2\sqrt{3}}{1 + 4 - 3} = \tan^{-1}(-\sqrt{3}) = 60°.$$ ∎

§ Problem 4.2.5. $(m^2 - mn)y = (mn + n^2)x + n^3$ and $(mn + m^2)y = (mn - n^2)x + m^3$. ◇

§§ Solution. The given equations are
$$(m^2 - mn)y = (mn + n^2)x + n^3,$$
$$\therefore y = \frac{mn + n^2}{m^2 - mn}x + \frac{n^3}{m^2 - mn},$$
$$\therefore m_1 = \frac{mn + n^2}{m^2 - mn}$$
$$(mn + m^2)y = (mn - n^2)x + m^3,$$
$$\therefore y = \frac{mn - n^2}{mn + m^2}x + \frac{m^3}{mn + m^2}, \therefore m_2 = \frac{mn - n^2}{mn + m^2}$$
Hence the angle between these two lines is
$$= \tan^{-1} \frac{m_1 - m_2}{1 + m_1 m_2} = \tan^{-1} \frac{\frac{mn + n^2}{m^2 - mn} - \frac{mn - n^2}{mn + m^2}}{1 + \frac{mn + n^2}{m^2 - mn} \cdot \frac{mn - n^2}{mn + m^2}}$$

Let us simplify the numerator and denominator separately.

$$\therefore \frac{mn + n^2}{m^2 - mn} - \frac{mn - n^2}{mn + m^2} = \frac{n(m + n)}{m(m - n)} - \frac{n(m - n)}{m(n + m)}$$
$$= \frac{n}{m} \left[\frac{(m + n)^2 - (m - n)^2}{m^2 - n^2} \right] = \frac{4n^2}{m^2 - n^2}$$
$$\therefore 1 + \frac{mn + n^2}{m^2 - mn} \cdot \frac{mn - n^2}{mn + m^2} = 1 + \frac{n(m + n)}{m(m - n)} \cdot \frac{n(m - n)}{m(n + m)}$$
$$= 1 + \frac{n^2}{m^2} = \frac{m^2 + n^2}{m^2}$$
Hence the angle between these two lines is
$$= \tan^{-1} \frac{\frac{4n^2}{m^2 - n^2}}{\frac{m^2 + n^2}{m^2}} = \tan^{-1} \frac{4m^2 n^2}{m^4 - n^4}.$$ ∎

§ Problem 4.2.6. *Find the tangent of the angle between the lines whose intercepts on the axes are respectively a, $-b$ and b, $-a$.* ◇

4.2. Angles between straight lines

§§ Solution. The equations to the lines are:
$$\frac{x}{a} - \frac{y}{b} = 1, \therefore y = \frac{b}{a}x - b, \therefore m_1 = \frac{b}{a}$$
$$\frac{x}{b} - \frac{y}{a} = 1, \therefore y = \frac{a}{b}x - a, \therefore m_2 = \frac{a}{b}$$

Hence the angle between these two lines is

$$= \tan^{-1} \frac{m_1 - m_2}{1 + m_1 m_2} = \tan^{-1} \frac{\frac{b}{a} - \frac{a}{b}}{1 + \frac{b}{a} \cdot \frac{a}{b}}$$

$$= \tan^{-1} \frac{b^2 - a^2}{2ab} = \tan^{-1} \frac{a^2 - b^2}{2ab}. \qquad \blacksquare$$

§ Problem 4.2.7. *Prove that the points* $(2, -1)$, $(0, 2)$, $(2, 3)$ *and* $(4, 0)$ *are the coordinates of the angular points of a parallelogram and find the angle between its diagonals.* ◊

§§ Solution. Let us denote the four points by $A(2, -1)$, $B(4, 0)$, $C(2, 3)$ and $D(0, 2)$ respectively.

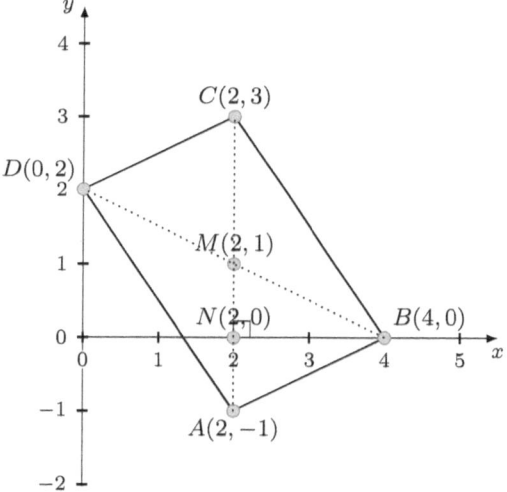

The coordinates of the middle point of the diagonal AC are
$$\left(\frac{2+2}{2}, \frac{-1+3}{2}\right) = (2, 1).$$

The coordinates of the middle point of the diagonal BD are
$$\left(\frac{4+0}{2}, \frac{0+2}{2}\right) = (2, 1).$$

Hence the middle point of each diagonal is the same, i.e., $M(2, 1)$. Therefore the diagonals of the quadrilateral $ABCD$ bisect each other. Hence it is a parallelogram.

Since the x-coordinate of the end points of the diagonal AC is the same, i.e., 2, hence AC is parallel to the y-axis, therefore AC is perpendicular to the x-axis.

4.2. Angles between straight lines

Slope of the diagonal $BD = \tan \angle DBX = \dfrac{2-0}{0-4} = -\dfrac{1}{2}$

$\therefore \angle DBX = \angle MBX = \tan^{-1}\left(-\dfrac{1}{2}\right)$

Hence the angle between the diagonals AC and BD is $\angle AMB = \angle NMB$

$\therefore \angle NMB = \pi - \angle MNB - \angle MBN$

$= \pi - \dfrac{\pi}{2} - (\pi - \angle MBX) = \angle MBX - \dfrac{\pi}{2}$

$= \tan^{-1}\left(-\dfrac{1}{2}\right) - \dfrac{\pi}{2} = \dfrac{\pi}{2} - \tan^{-1}(-2) - \dfrac{\pi}{2} = \tan^{-1} 2.$ ∎

Find the equation to the straight line

§ Problem 4.2.8. *passing through the point $(2,3)$ and perpendicular to the straight line $4x - 3y = 10$.* ◊

§§ Solution. The equation to the straight line perpendicular to the straight line $4x - 3y = 10$ is
$$3x + 4y = c$$
Since it passes through the point $(2,3)$,
$$\therefore 3 \cdot 2 + 4 \cdot 3 = c$$
$$\therefore c = 18$$
Hence the equation to the straight line is $3x + 4y = 18$.

Alternative Solution :

Let us find the slope of the given line
$$4x - 3y = 10$$
$$\therefore y = \dfrac{4}{3}x - \dfrac{10}{3}$$
Hence its slope is $\dfrac{4}{3}$.

Hence slope of the straight line perpendicular to the above is $m = -\dfrac{3}{4}$.

The equation to the straight line passing through the point $(2,3)$ is
$$y - 3 = m(x - 2) = -\dfrac{3}{4}(x - 2)$$
$$\therefore 4y - 12 = -3x + 6$$
$$\therefore 3x + 4y = 18.$$ ∎

§ Problem 4.2.9. *passing through the point $(-6, 10)$ and perpendicular to the straight line $7x + 8y = 5$.* ◊

§§ Solution. The equation to the straight line perpendicular to the straight line $7x + 8y = 5$ is
$$8x - 7y = c$$
Since it passes through the point $(-6, 10)$,
$$\therefore 8 \cdot (-6) - 7 \cdot 10 = c$$
$$\therefore c = -118$$
Hence the equation to the straight line is
$$8x - 7y = -118$$
$$\therefore 7y - 8x = 118.$$ ∎

4.2. Angles between straight lines

§ Problem 4.2.10. *passing through the point* $(2, -3)$ *and perpendicular to the straight line joining the points* $(5, 7)$ *and* $(-6, 3)$. ◊

§§ Solution. Slope of the line joining the points $(5, 7)$ and $(-6, 3)$ is
$$\frac{7-3}{5-(-6)} = \frac{4}{11}.$$

Hence slope of the line perpendicular to this line is $m = -\frac{11}{4}$.

Hence the equation of the straight line with slope $m = -\frac{11}{4}$ and passing through the point $(2, -3)$ is
$$y - (-3) = -\frac{11}{4}(x - 2)$$
$$\therefore 4y + 12 = 22 - 11x$$
$$\therefore 4y + 11x = 10. \quad \blacksquare$$

§ Problem 4.2.11. *passing through the point* $(-4, -3)$ *and perpendicular to the straight line joining* $(1, 3)$ *and* $(2, 7)$. ◊

§§ Solution. Slope of the line joining the points $(1, 3)$ and $(2, 7)$ is
$$\frac{7-3}{2-1} = 4.$$

Hence slope of the line perpendicular to this line is $m = -\frac{1}{4}$.

Hence the equation of the straight line with slope $m = -\frac{1}{4}$ and passing through the point $(-4, -3)$ is
$$y + 3 = -\frac{1}{4}(x + 4)$$
$$\therefore 4y + 12 = -x - 4$$
$$\therefore x + 4y + 16 = 0. \quad \blacksquare$$

§ Problem 4.2.12. *Find the equation to the straight line drawn at right angles to the straight line* $\frac{x}{a} - \frac{y}{b} = 1$ *through the point where it meets the axis of* x. ◊

§§ Solution. Equation of the straight line perpendicular to the line $\frac{x}{a} - \frac{y}{b} = 1$ is $ax + by = c$.

Since it passes through the point $(a, 0)$, hence $c = a^2$.

Hence the equation is $ax + by = a^2$. $\quad \blacksquare$

§ Problem 4.2.13. *Find the equation to the straight line which bisects, and is perpendicular to, the straight line joining the points* (a, b) *and* (a', b'). ◊

§§ Solution. The slope of the line joining the points (a, b) and (a', b') is $\frac{b - b'}{a - a'}$.

Hence slope of the line perpendicular to this line is $-\frac{a - a'}{b - b'}$.

The coordinates of the point bisecting the straight line joining the points (a, b) and (a', b') are $\left(\frac{a + a'}{2}, \frac{b + b'}{2}\right)$.

Hence the required equation is
$$y - \frac{b + b'}{2} = -\frac{a - a'}{b - b'}\left(x - \frac{a + a'}{2}\right)$$

4.2. Angles between straight lines

$$\therefore 2y(b-b') - b^2 + b'^2 = -2x(a-a') + a^2 - a'^2$$
$$\therefore 2(a-a')x + 2(b-b')y = a^2 - a'^2 + b^2 - b'^2.\qquad\blacksquare$$

§ Problem 4.2.14. *Prove that the equation to the straight line which passes through the point $\left(a\cos^3\theta,\ a\sin^3\theta\right)$ and is perpendicular to the straight line $x\sec\theta + y\operatorname{cosec}\theta = a$ is $x\cos\theta - y\sin\theta = a\cos 2\theta$.* ◊

§§ Solution. Equation of the straight line perpendicular to the line $x\sec\theta + y\ \operatorname{cosec}\theta = a$ is
$$x\cos\theta - y\sin\theta = c.$$
Since this line passes through the point $(a\cos^3\theta, a\sin^3\theta)$,
$$\therefore a\cos^4\theta - a\sin^4\theta = c$$
$$\therefore c = a(\cos^2\theta + \sin^2\theta)(\cos^2\theta - \sin^2\theta) = a\cos 2\theta.$$
Hence the required equation is
$$x\cos\theta - y\sin\theta = a\cos 2\theta.\qquad\blacksquare$$

§ Problem 4.2.15. *Find the equations to the straight lines passing through (x', y') and respectively perpendicular to the straight lines*
$$xx' + yy' = a^2,$$
$$\frac{xx'}{a^2} + \frac{yy'}{b^2} = 1,$$
and $x'y + xy' = a^2$.
◊

§§ Solution. The equations to the straight lines perpendicular to the given lines are
$$xy' - x'y = c_1$$
$$a^2 y'x - b^2 x'y = c_2$$
$$xx' - yy' = c_3$$
Sine these lines passes through the point (x', y'), we have
$$x'y' - x'y' = c_1,\ \therefore c_1 = 0.$$
$$a^2 y'x' - b^2 x'y' = c_2,\ \therefore c_2 = (a^2 - b^2)x'y'.$$
$$x'x' - y'y' = c_3,\ \therefore c_3 = x'^2 - y'^2.$$
Hence the required equations are
$$xy' - x'y = 0$$
$$a^2 y'x - b^2 x'y = (a^2 - b^2)x'y'$$
$$xx' - yy' = x'^2 - y'^2\qquad\blacksquare$$

§ Problem 4.2.16. *Find the equations to the straight Hues which divide, internally and externally, the line joining $(-3, 7)$ to $(5, -4)$ in the ratio of $4:7$ and which are perpendicular to this line.* ◊

§§ Solution. The slope of the line joining $(-3, 7)$ to $(5, -4)$ is
$$\frac{7-(-4)}{-3-5} = -\frac{11}{8}$$
Hence slope of the lines perpendicular to the above line is $=\dfrac{8}{11}$.

The lines pass through the points
$$\left(\frac{4\cdot 5 - 3\cdot 7}{4+7},\ \frac{-4\cdot 4 + 7\cdot 7}{4+7}\right);\ \left(\frac{4\cdot 5 + 3\cdot 7}{4-7},\ \frac{-4\cdot 4 - 7\cdot 7}{4-7}\right)$$
i.e.
$$\left(-\frac{1}{11}, 3\right);\ \left(-\frac{41}{3}, \frac{65}{3}\right)$$

4.2. Angles between straight lines

Hence the required equations are
$$y - 3 = \frac{8}{11}\left(x + \frac{1}{11}\right), \text{ and } y - \frac{65}{3} = \frac{8}{11}\left(x + \frac{41}{3}\right)$$
i.e. $121y - 88x = 371$, and $33y - 24x = 1043$. ∎

§ Problem 4.2.17. *Through the point $(3, 4)$ are drawn two straight lines each inclined at $45°$ to the straight line $x - y = 2$. Find their equations and find also the area included by the three lines.* ◊

§§ Solution. The slope m of the line
$$x - y = 2 \tag{4.16}$$
is $m = 1$.

By Art. 72, the required equations are
$$y - 4 = \frac{1 + \tan 45°}{1 - 1 \cdot \tan 45°}(x - 3), \text{ and}$$
$$y - 4 = \frac{1 - \tan 45°}{1 + 1 \cdot \tan 45°}(x - 3)$$

i.e.
$$y - 4 = \frac{1 + 1}{1 - 1}(x - 3), \text{ and}$$
$$y - 4 = \frac{1 - 1}{1 + 1}(x - 3)$$

i.e.
$$x - 3 = 0, \text{ and} \tag{4.17}$$
$$y - 4 = 0. \tag{4.18}$$

Let us draw the lines as follows:

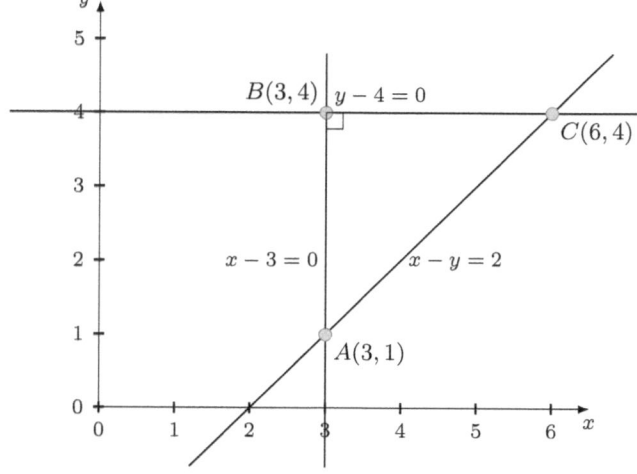

Solving (4.16) and (4.17), we get the coordinates of A as $(3, 1)$.
Solving (4.16) and (4.18), we get the coordinates of C as $(6, 4)$.
Solving (4.17) and (4.18), we get the coordinates of B as $(3, 4)$.

Hence ABC is a right-angled isosceles triangle whose equal sides $AB = BC = 3$.

4.2. Angles between straight lines

Hence the area of the triangle $ABC = \frac{1}{2} \times 3 \times 3 = 4\frac{1}{2}$. ∎

§ Problem 4.2.18. *Show that the equations to the straight lines passing through the point $(3, -2)$ and inclined at $60°$ to the line $\sqrt{3}x + y = 1$ are $y + 2 = 0$ and $y - \sqrt{3}x + 2 + 3\sqrt{3} = 0$.*

◊

§§ Solution. The slope of the line $\sqrt{3}x + y = 1$ is $m = -\sqrt{3}$.
Hence by Art. 72, the required equations are
$$y - (-2) = \frac{-\sqrt{3} + \tan 60°}{1 - (-\sqrt{3})\tan 60°}(x - 3), \text{ and}$$
$$y - (-2) = \frac{-\sqrt{3} - \tan 60°}{1 + (-\sqrt{3})\tan 60°}(x - 3)$$

i.e.
$$y + 2 = \frac{-\sqrt{3} + \sqrt{3}}{1 + 3}(x - 3), \text{ and}$$
$$y + 2 = \frac{-\sqrt{3} - \sqrt{3}}{1 - 3}(x - 3)$$

i.e.
$$y + 2 = 0, \text{ and}$$
$$y - \sqrt{3}x + 2 + 3\sqrt{3} = 0.$$
∎

§ Problem 4.2.19. *Find the equations to the straight lines which pass through the origin and are inclined at $75°$ to the straight line $x + y + \sqrt{3}(y - x) = a$.*

◊

§§ Solution. The slope of the line $x + y + \sqrt{3}(y - x) = a$ is $m = \dfrac{\sqrt{3} - 1}{\sqrt{3} + 1}$.

Hence by Art. 72, the required equations are
$$y - 0 = \frac{\dfrac{\sqrt{3}-1}{\sqrt{3}+1} + \tan 75°}{1 - \dfrac{\sqrt{3}-1}{\sqrt{3}+1}\tan 75°}(x - 0), \text{ and },$$

$$y - 0 = \frac{\dfrac{\sqrt{3}-1}{\sqrt{3}+1} - \tan 75°}{1 + \dfrac{\sqrt{3}-1}{\sqrt{3}+1}\tan 75°}(x - 0)$$

∵ $\tan 75° = \dfrac{\sqrt{3}+1}{\sqrt{3}-1}$

$$y = \frac{\dfrac{\sqrt{3}-1}{\sqrt{3}+1} + \dfrac{\sqrt{3}+1}{\sqrt{3}-1}}{1 - \dfrac{\sqrt{3}-1}{\sqrt{3}+1}\dfrac{\sqrt{3}+1}{\sqrt{3}-1}}x, \text{ and,}$$

$$y = \frac{\dfrac{\sqrt{3}-1}{\sqrt{3}+1} - \dfrac{\sqrt{3}+1}{\sqrt{3}-1}}{1 + \dfrac{\sqrt{3}-1}{\sqrt{3}+1}\dfrac{\sqrt{3}+1}{\sqrt{3}-1}}x$$

4.2. Angles between straight lines 64

i.e.
$$x = 0, \text{ and}$$
$$y = \frac{(\sqrt{3}-1)^2 - (\sqrt{3}+1)^2}{1+1} \cdot \frac{3-1}{} = -\sqrt{3}x.$$

Hence the required equations are
$$x = 0,$$
$$y + \sqrt{3}x = 0.$$ ∎

§ Problem 4.2.20. *Find the equations to the straight lines which pass through the point (h, k) and are inclined at an angle $\tan^{-1} m$ to the straight line $y = mx + c$.* ◊

§§ Solution. By Art. 72, the required equations are
$$y - k = \frac{m + m}{1 - m^2}(x - h), \text{ and}$$
$$y - k = \frac{m - m}{1 + m^2}(x - h)$$

i.e.
$$(1 - m^2)(y - k) = 2m(x - h), \text{ and}$$
$$y = k.$$ ∎

§ Problem 4.2.21. *Find the angle between the two straight lines $3x = 4y + 7$ and $5y = 12x + 6$ and also the equations to the two straight lines which pass through the point $(4, 5)$ and make equal angles with the two given lines.* ◊

§§ Solution. The slope of the line $5y = 12x + 6$ is $m_1 = \frac{12}{5}$.

The slope of the line $3x = 4y + 7$ is $m_2 = \frac{3}{4}$.

Hence the angle between these two lines is
$$= \tan^{-1} \frac{m_1 - m_2}{1 + m_1 m_2} = \tan^{-1} \frac{\frac{12}{5} - \frac{3}{4}}{1 + \frac{12}{5} \cdot \frac{3}{4}} = \tan^{-1} \frac{33}{56}.$$

Let us draw the lines.

Let θ_1 and θ_2 be the angles made by the lines $3x = 4y + 7$ and $5y = 12x + 6$ with the x-axis respectively.

It is clear from the picture that $\tan \theta_1 = \frac{3}{4}$ and $\tan \theta_2 = \frac{12}{5}$.

Let θ be the angle made by the dashed line with the x-axis, which makes equal angles α with the two given lines.
$$\therefore \angle PAB = \alpha$$
$$\text{and } \angle PAC = \alpha.$$
$$\therefore \angle CAB = 2\alpha.$$

From the triangle ABP,
$$\angle BAP + \angle APB + \angle PBA = 180°$$
$$\therefore \alpha + \theta + 180° - \theta_2 = 180°$$

$$\therefore \alpha = \theta_2 - \theta. \tag{4.19}$$

4.2. Angles between straight lines

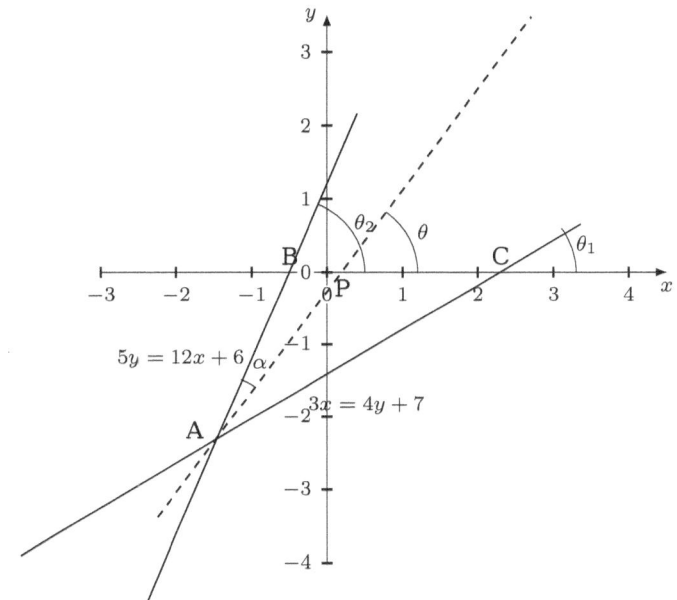

From the triangle PAC,
$$\angle PAC + \angle ACP + \angle CPA = 180°$$
$$\therefore \alpha + \theta_1 + 180° - \theta = 180°$$

$$\therefore \alpha = \theta - \theta_1. \tag{4.20}$$

From (4.19) and (4.20), we get
$$\theta_2 - \theta = \theta - \theta_1$$
$$\therefore 2\theta = \theta_1 + \theta_2.$$
$$\therefore \tan 2\theta = \tan(\theta_1 + \theta_2)$$

$$\therefore \frac{2\tan\theta}{1-\tan^2\theta} = \frac{\tan\theta_1 + \tan\theta_2}{1 - \tan\theta_1 \cdot \tan\theta_2} = \frac{\frac{3}{4} + \frac{12}{5}}{1 - \frac{3}{4} \cdot \frac{12}{5}} = -\frac{63}{16}$$

$$\therefore 63\tan^2\theta - 32\tan\theta - 63 = 0$$
$$\therefore (7\tan\theta - 9)(9\tan\theta + 7) = 0$$
$$\therefore \tan\theta = \frac{9}{7}, \text{ or } -\frac{7}{9}.$$

Hence the required equations are
$$y - 5 = \frac{9}{7}(x - 4), \text{ and}$$
$$y - 5 = -\frac{7}{9}(x - 4)$$

i.e.
$$9x - 7y = 1, \text{ and}$$
$$7x + 9y = 73.$$

4.3. Lengths of Perpendiculars

Alternative Solution : Let the equation of the line passing through the point $(4, 5)$ be
$$y - 5 = m(x - 4) \tag{4.21}$$
Since this line makes equal angles with the lines $3x = 4y + 7$ and $5y = 12x + 6$

$$\therefore \frac{m - \frac{3}{4}}{1 + m \cdot \frac{3}{4}} = \frac{\frac{12}{5} - m}{1 + \frac{12}{5} \cdot m}$$

$$\therefore \frac{4m - 3}{4 + 3m} = \frac{12 - 5m}{5 + 12m}$$

$$\therefore 20m + 48m^2 - 15 - 36m = 48 + 36m - 20m - 15m^2$$

$$\therefore 63m^2 - 32m - 63 = 0$$

$$\therefore (7m - 9)(9m + 7) = 0$$

$$\therefore m = \frac{9}{7} \text{ or } -\frac{7}{9}.$$

Putting these values of m in (4.21), the required equations are:
$$y - 5 = \frac{9}{7}(x - 4), \text{ and}$$
$$y - 5 = -\frac{7}{9}(x - 4)$$

i.e.
$$9x - 7y = 1, \text{ and}$$
$$7x + 9y = 73.$$ ∎

4.3 Lengths of Perpendiculars

Find the length of the perpendicular drawn from

§ Problem 4.3.1. *the point $(4, 5)$ upon the straight line $3x + 4y = 10$.* ◊

§§ Solution. The length of the perpendicular
$$= \frac{3 \cdot 4 + 4 \cdot 5 - 10}{\sqrt{3^2 + 4^2}} = \frac{22}{5} = 4\frac{2}{5}.$$ ∎

§ Problem 4.3.2. *the origin upon the straight line $\frac{x}{3} - \frac{y}{4} = 1$.* ◊

§§ Solution. The length of the perpendicular
$$= \frac{1}{\sqrt{\frac{1}{3^2} + \frac{1}{4^2}}} = \frac{12}{5} = 2\frac{2}{5}.$$ ∎

§ Problem 4.3.3. *the point $(-3, -4)$ upon the straight line*
$$12(x + 6) = 5(y - 2).$$ ◊

§§ Solution. The equation to the straight line is
$$12(x + 6) = 5(y - 2)$$
$$\therefore 12x - 5y + 82 = 0.$$

Hence the length of the perpendicular
$$= \frac{12 \cdot (-3) - 5 \cdot (-4) + 82}{\sqrt{12^2 + 5^2}} = \frac{66}{13} = 5\frac{1}{13}.$$ ∎

§ Problem 4.3.4. *the point (b, a) upon the straight line*
$$\frac{x}{a} - \frac{y}{b} = 1.$$ ◊

4.3. Lengths of Perpendiculars

§§ Solution. The length of the perpendicular

$$= \frac{\frac{b}{a} - \frac{a}{b} - 1}{\sqrt{\frac{1}{a^2} + \frac{1}{b^2}}} = \frac{b^2 - a^2 - ab}{\sqrt{a^2 + b^2}}.$$ ∎

§ Problem 4.3.5. *Find the length of the perpendicular from the origin upon the straight line joining the two points whose coordinates are*

$$(a\cos\alpha, a\sin\alpha) \text{ and } (a\cos\beta, a\sin\beta).$$ ◊

§§ Solution. The equation to the straight line is already found in Example v : Question 22.

Let us redo this exercise of finding equation here for the sake of simplicity.

The equation to the straight line is :

$$y - a\sin\alpha = \frac{a\sin\beta - a\sin\alpha}{a\cos\beta - a\cos\alpha}(x - a\cos\alpha)$$

$$\therefore y - a\sin\alpha = \frac{2\cos\frac{\beta+\alpha}{2}\sin\frac{\beta-\alpha}{2}}{-2\sin\frac{\beta+\alpha}{2}\sin\frac{\beta-\alpha}{2}}(x - a\cos\alpha)$$

$$\therefore y\sin\frac{\alpha+\beta}{2} - a\sin\alpha\sin\frac{\alpha+\beta}{2} = a\cos\alpha\cos\frac{\alpha+\beta}{2} - x\cos\frac{\alpha+\beta}{2}$$

$$\therefore x\cos\frac{\alpha+\beta}{2} + y\sin\frac{\alpha+\beta}{2} = a\cos\left(\alpha - \frac{\alpha+\beta}{2}\right)$$

$$\therefore x\cos\frac{\alpha+\beta}{2} + y\sin\frac{\alpha+\beta}{2} = a\cos\frac{\alpha-\beta}{2}. \tag{4.22}$$

We already know that the equation of the following form makes an angle α with the x-axis and the length of the perpendicular on it from the origin is p:

$$x\cos\alpha + y\sin\alpha - p = 0.$$

Hence (4.22) makes an angle $\frac{\alpha+\beta}{2}$ with the x-axis and the length of the perpendicular on it from the origin is $a\cos\frac{\alpha-\beta}{2}$.

Alternatively, the length of the perpendicular on it from the origin $(0,0)$ is

$$= \frac{0 \times \cos\frac{\alpha+\beta}{2} + 0 \times \sin\frac{\alpha+\beta}{2} - a\cos\frac{\alpha-\beta}{2}}{\sqrt{\cos^2\frac{\alpha+\beta}{2} + \sin^2\frac{\alpha+\beta}{2}}}$$

$$= a\cos\frac{\alpha-\beta}{2}. \text{ (ignoring the -ve sign.)}$$ ∎

§ Problem 4.3.6. *Show that the product of the perpendiculars drawn from the two points $\left(\pm\sqrt{a^2 - b^2}, 0\right)$ upon the straight line*

$$\frac{x}{a}\cos\theta + \frac{y}{b}\sin\theta = 1 \text{ is } b^2.$$ ◊

§§ Solution. Product of the perpendiculars

$$= \frac{\frac{\sqrt{a^2-b^2}}{a}\cos\theta + 0 - 1}{\sqrt{\frac{\cos^2\theta}{a^2} + \frac{\sin^2\theta}{b^2}}} \times \frac{-\frac{\sqrt{a^2-b^2}}{a}\cos\theta + 0 - 1}{\sqrt{\frac{\cos^2\theta}{a^2} + \frac{\sin^2\theta}{b^2}}}$$

4.3. Lengths of Perpendiculars

$$= \frac{b\sqrt{a^2-b^2}\cos\theta - ab}{\sqrt{b^2\cos^2\theta + a^2\sin^2\theta}} \times \frac{-b\sqrt{a^2-b^2}\cos\theta - ab}{\sqrt{b^2\cos^2\theta + a^2\sin^2\theta}}$$

$$= \frac{a^2b^2 - b^2(a^2-b^2)\cos^2\theta}{b^2\cos^2\theta + a^2\sin^2\theta}$$

$$= \frac{b^2(a^2 - a^2\cos^2\theta + b^2\cos^2\theta)}{b^2\cos^2\theta + a^2\sin^2\theta}$$

$$= \frac{b^2(a^2\sin^2\theta + b^2\cos^2\theta)}{b^2\cos^2\theta + a^2\sin^2\theta}$$

$$= b^2. \qquad \blacksquare$$

§ Problem 4.3.7. *If p and p' be the perpendiculars from the origin upon the straight lines whose equations are*

$$x\sec\theta + y\csc\theta = a, \text{ and}$$
$$x\cos\theta - y\sin\theta = a\cos 2\theta,$$

prove that
$$4p^2 + p'^2 = a^2. \qquad \diamond$$

§§ Solution.
$$p = \frac{a}{\sqrt{\sec^2\theta + \csc^2\theta}} = a\sin\theta\cdot\cos\theta = \frac{1}{2}a\sin 2\theta.$$
$$p' = a\cos 2\theta.$$

$$4p^2 + p'^2 = a^2(\sin^2 2\theta + \cos^2 2\theta) = a^2. \qquad \blacksquare$$

§ Problem 4.3.8. *Find the distance between the two parallel straight lines*
$$y = mx + c, \text{ and}$$
$$y = mx + d. \qquad \diamond$$

§§ Solution. Length of the perpendicular from the origin on the line $y = mx + c$ is

$$\frac{c}{\sqrt{1+m^2}}$$

Length of the perpendicular from the origin on the line $y = mx + d$ is

$$\frac{d}{\sqrt{1+m^2}}$$

Hence the distance between these two parallel lines is

$$\frac{c-d}{\sqrt{1+m^2}}$$

Alternative Solution :

The distance between these two parallel lines is equal to the length of the perpendicular from a given point on first line upon second line.

It is easy to see that the point $(0, c)$ lies on the line $y = mx + c$.

$\therefore$ Length of the perpendicular from this point $(0, c)$ upon the line $y = mx + d$ is

$$\frac{c-d}{\sqrt{1+m^2}}.$$

Hence the distance between these two parallel lines is

$$\frac{c-d}{\sqrt{1+m^2}} \qquad \blacksquare$$

4.4. Bisectors of angles between straight lines

§ Problem 4.3.9. *What are the points on the axis of x whose perpendicular distance from the straight line $\frac{x}{a} + \frac{y}{b} = 1$ is a?* ◊

§§ Solution. The perpendicular distance from a given point $(h, 0)$ on the x-axis upon the given straight line is

$$= \frac{\frac{h}{a} + 0 - 1}{\sqrt{\frac{1}{a^2} + \frac{1}{b^2}}} = \frac{bh - ab}{\sqrt{a^2 + b^2}}$$

$$\therefore \frac{bh - ab}{\sqrt{a^2 + b^2}} = \pm a$$

$$\therefore h = \frac{a}{b}\left(b \pm \sqrt{a^2 + b^2}\right).$$

Hence the required points are $\left\{\frac{a}{b}\left(b \pm \sqrt{a^2 + b^2}\right), 0\right\}$. ∎

§ Problem 4.3.10. *Show that the perpendiculars let fall from any point of the straight line $2x + 11y = 5$ upon the two straight lines $24x + 7y = 20$ and $4x - 3y = 2$ are equal to each other.* ◊

§§ Solution. Let (h, k) be any point such that the perpendiculars from it upon the given lines are equal.

$$\therefore \frac{24h + 7k - 20}{\sqrt{24^2 + 7^2}} = \frac{4h - 3k - 2}{\sqrt{4^2 + 3^2}}$$

$$\therefore 24h + 7k - 20 = 5(4h - 3k - 2)$$

$$\therefore 2h + 11k = 5.$$

Hence it is clear that if the point (h, k) lies on the straight line $2x + 11y = 5$, the perpendiculars are equal. ∎

§ Problem 4.3.11. *Find the perpendicular distance from the origin of the perpendicular from the point $(1, 2)$ upon the straight line*

$$x - \sqrt{3}y + 4 = 0.$$ ◊

§§ Solution. Slope of the line $x - \sqrt{3}y + 4 = 0$ is $\frac{1}{\sqrt{3}}$.

Hence slope of the line perpendicular to the above line is $-\sqrt{3}$. Since this passes through the point $(1, 2)$, its equation is

$$y - 2 = -\sqrt{3}(x - 1)$$

$$\therefore \sqrt{3}x + y - 2 - \sqrt{3} = 0.$$

Hence the length of the perpendicular on it from the origin $(0, 0)$ is

$$= \frac{2 + \sqrt{3}}{\sqrt{4}} = \frac{2 + \sqrt{3}}{2}.$$ ∎

4.4 Bisectors of angles between straight lines

Find the coordinates of the points of intersection of the straight lines whose equations are

§ Problem 4.4.1. $2x - 3y + 5 = 0$ and $7x + 4y = 3$. ◊

§§ Solution. The coordinates of the points of intersection are given by (x, y):

$$\frac{x}{b_1 c_2 - b_2 c_1} = \frac{y}{c_1 a_2 - c_2 a_1} = \frac{1}{a_1 b_2 - a_2 b_1}$$

4.4. Bisectors of angles between straight lines

$$\therefore \frac{x}{-3 \times -3 - 4 \times 5} = \frac{y}{5 \times 7 - (-3) \times 2} = \frac{1}{2 \times 4 - 7 \times (-3)}$$

$$\therefore \frac{x}{-11} = \frac{y}{41} = \frac{1}{29}$$

$$\therefore x = -\frac{11}{29}, \ y = \frac{41}{29}.$$

Hence the required point is $\left(-\dfrac{11}{29}, \dfrac{41}{29}\right)$ ∎

§ Problem 4.4.2. $\dfrac{x}{a} + \dfrac{y}{b} = 1$ and $\dfrac{x}{b} + \dfrac{y}{a} = 1$. ◊

§§ Solution. The coordinates of the points of intersection are given by (x, y):

$$\frac{x}{b_1 c_2 - b_2 c_1} = \frac{y}{c_1 a_2 - c_2 a_1} = \frac{1}{a_1 b_2 - a_2 b_1}$$

$$\therefore \frac{x}{\dfrac{1}{b} \times (-1) - \dfrac{1}{a} \times (-1)} = \frac{y}{-1 \times \dfrac{1}{b} - (-1) \times \dfrac{1}{a}} = \frac{1}{\dfrac{1}{a} \times \dfrac{1}{a} - \dfrac{1}{b} \times \dfrac{1}{b}}$$

$$\therefore \frac{x}{\dfrac{1}{a} - \dfrac{1}{b}} = \frac{y}{\dfrac{1}{a} - \dfrac{1}{b}} = \frac{1}{\left(\dfrac{1}{a} - \dfrac{1}{b}\right)\left(\dfrac{1}{a} + \dfrac{1}{b}\right)}$$

$$\therefore x = y = \frac{ab}{a+b}$$

Hence the required point is $\left(\dfrac{ab}{a+b}, \dfrac{ab}{a+b}\right)$ ∎

§ Problem 4.4.3. $y = m_1 x + \dfrac{a}{m_1}$ and $y = m_2 x + \dfrac{a}{m_2}$. ◊

§§ Solution. The coordinates of the points of intersection are given by (x, y):

$$\frac{x}{b_1 c_2 - b_2 c_1} = \frac{y}{c_1 a_2 - c_2 a_1} = \frac{1}{a_1 b_2 - a_2 b_1}$$

$$\therefore \frac{x}{(-1) \times \dfrac{a}{m_2} - (-1) \times \dfrac{a}{m_1}} = \frac{y}{\dfrac{a}{m_1} \times m_2 - \dfrac{a}{m_2} \times m_1}$$

$$= \frac{1}{m_1 \times (-1) - m_2 \times (-1)}$$

$$\therefore \frac{x}{a\left(\dfrac{1}{m_1} - \dfrac{1}{m_2}\right)} = \frac{y}{a\left(\dfrac{m_2}{m_1} - \dfrac{m_1}{m_2}\right)} = \frac{1}{m_2 - m_1}$$

$$\therefore \frac{x}{\dfrac{a}{m_1 m_2}(m_2 - m_1)} = \frac{y}{\dfrac{a}{m_1 m_2}(m_2^2 - m_1^2)} = \frac{1}{m_2 - m_1}$$

$$\therefore x = \frac{a}{m_1 m_2}, \ y = a\left\{\frac{1}{m_1} + \frac{1}{m_2}\right\}$$

Hence the required point is $\left(\dfrac{a}{m_1 m_2}, \ a\left\{\dfrac{1}{m_1} + \dfrac{1}{m_2}\right\}\right)$ ∎

§ Problem 4.4.4. $x \cos\phi_1 + y \sin\phi_1 = a$ and $x \cos\phi_2 + y \sin\phi_2 = a$. ◊

§§ Solution. The coordinates of the points of intersection are given by (x, y):

$$\frac{x}{b_1 c_2 - b_2 c_1} = \frac{y}{c_1 a_2 - c_2 a_1} = \frac{1}{a_1 b_2 - a_2 b_1}$$

$$\therefore \frac{x}{\sin\phi_1 \times (-a) - \sin\phi_2 \times (-a)} = \frac{y}{(-a) \times \cos\phi_2 - (-a) \times \cos\phi_1}$$

4.4. Bisectors of angles between straight lines

$$= \frac{1}{\cos\phi_1 \times \sin\phi_2 - \cos\phi_2 \times \sin\phi_1}$$

$$\therefore \frac{x}{\sin\phi_2 - \sin\phi_1} = \frac{y}{\cos\phi_1 - \cos\phi_2} = \frac{a}{\sin(\phi_2 - \phi_1)}$$

$$\therefore \frac{x}{2\cos\frac{\phi_2+\phi_1}{2}\sin\frac{\phi_2-\phi_1}{2}} = \frac{y}{2\sin\frac{\phi_1+\phi_2}{2}\sin\frac{\phi_2-\phi_1}{2}}$$

$$= \frac{a}{2\sin\frac{\phi_2-\phi_1}{2}\cos\frac{\phi_2-\phi_1}{2}}$$

$$\therefore x = a\frac{\cos\frac{1}{2}(\phi_1+\phi_2)}{\cos\frac{1}{2}(\phi_1-\phi_2)}, \quad y = a\frac{\sin\frac{1}{2}(\phi_1+\phi_2)}{\cos\frac{1}{2}(\phi_1-\phi_2)}.$$

Hence the required point is
$$\left\{a\cos\frac{1}{2}(\phi_1+\phi_2)\sec\frac{1}{2}(\phi_1-\phi_2),\ a\sin\frac{1}{2}(\phi_1+\phi_2)\sec\frac{1}{2}(\phi_1-\phi_2)\right\}. \blacksquare$$

§ Problem 4.4.5. *Two straight lines cut the axis of x at distances a and $-a$ and the axis of y at distances b and b' respectively; find the coordinates of their point of intersection.* ◊

§§ Solution. For the first line, intercept on x-axis is a and intercept on y-axis is b, hence its equation is
$$\frac{x}{a} + \frac{y}{b} = 1$$

For the second line, intercept on x-axis is $-a$ and intercept on y-axis is b', hence its equation is
$$\frac{x}{-a} + \frac{y}{b'} = 1$$

The coordinates of the points of intersection are given by (x, y):
$$\frac{x}{b_1c_2 - b_2c_1} = \frac{y}{c_1a_2 - c_2a_1} = \frac{1}{a_1b_2 - a_2b_1}$$

$$\therefore \frac{x}{\frac{1}{b}\times(-1) - \frac{1}{b'}\times(-1)} = \frac{y}{(-1)\times\frac{1}{-a} - (-1)\times\frac{1}{a}} = \frac{1}{\frac{1}{a}\times\frac{1}{b'} - \frac{1}{-a}\times\frac{1}{b}}$$

$$\therefore \frac{x}{\frac{1}{b'} - \frac{1}{b}} = \frac{y}{\frac{2}{a}} = \frac{1}{\frac{1}{a}\left(\frac{1}{b'} + \frac{1}{b}\right)}$$

$$\therefore x = \frac{a(b-b')}{b+b'},\ y = \frac{2bb'}{b+b'}$$

Hence the required point is $\left(\dfrac{a(b-b')}{b+b'},\ \dfrac{2bb'}{b+b'}\right).$ $\blacksquare$

§ Problem 4.4.6. *Find the distance of the point of intersection of the two straight lines*
$$2x - 3y + 5 = 0 \text{ and } 3x + 4y = 0$$
from the straight line
$$5x - 2y = 0.$$
◊

§§ Solution. The coordinates of the points of intersection are given by (x, y):
$$\frac{x}{b_1c_2 - b_2c_1} = \frac{y}{c_1a_2 - c_2a_1} = \frac{1}{a_1b_2 - a_2b_1}$$

$$\therefore \frac{x}{(-3)\cdot 0 - 4\cdot 5} = \frac{y}{5\cdot 3 - 0\cdot 2} = \frac{1}{2\cdot 4 - 3\cdot(-3)}$$

4.4. Bisectors of angles between straight lines

$$\therefore x = -\frac{20}{17}, \; y = \frac{15}{17}.$$

Hence the point of intersection is $\left(-\frac{20}{17}, \frac{15}{17}\right)$.

Its distance from the line $5x - 2y = 0$ is the length of the perpendicular from this point to the line, i.e.

$$= \frac{5 \cdot -\frac{20}{17} - 2 \cdot \frac{15}{17}}{\sqrt{5^2 + 2^2}} = \frac{130}{17\sqrt{29}} \text{ (ignoring the -ve sign)} \quad \blacksquare$$

§ Problem 4.4.7. *Show that the perpendicular from the origin upon the straight line joining the points*

$$(a\cos\alpha, a\sin\alpha) \text{ and } (a\cos\beta, a\sin\beta)$$

bisects the distance between them. ◊

§§ Solution. The equation to the straight line is already found in § Problem 4.1.22.

Let us redo this exercise of finding equation here for the sake of simplicity.

The equation to the straight line is :

$$y - a\sin\alpha = \frac{a\sin\beta - a\sin\alpha}{a\cos\beta - a\cos\alpha}(x - a\cos\alpha)$$

$$\therefore y - a\sin\alpha = \frac{2\cos\frac{\beta+\alpha}{2}\sin\frac{\beta-\alpha}{2}}{-2\sin\frac{\beta+\alpha}{2}\sin\frac{\beta-\alpha}{2}}(x - a\cos\alpha)$$

$$\therefore y\sin\frac{\alpha+\beta}{2} - a\sin\alpha\sin\frac{\alpha+\beta}{2} = a\cos\alpha\cos\frac{\alpha+\beta}{2} - x\cos\frac{\alpha+\beta}{2}$$

$$\therefore x\cos\frac{\alpha+\beta}{2} + y\sin\frac{\alpha+\beta}{2} = a\cos\left(\alpha - \frac{\alpha+\beta}{2}\right)$$

$$\therefore x\cos\frac{\alpha+\beta}{2} + y\sin\frac{\alpha+\beta}{2} = a\cos\frac{\alpha-\beta}{2}. \quad (4.23)$$

By Art. 70, the equation to the line perpendicular to this line is

$$x\sin\frac{\alpha+\beta}{2} - y\cos\frac{\alpha+\beta}{2} + c = 0$$

Since it passes through the origin $(0,0)$, $\therefore c = 0$.

Hence the equation to the line perpendicular to (4.23) is

$$x\sin\frac{\alpha+\beta}{2} - y\cos\frac{\alpha+\beta}{2} = 0 \quad (4.24)$$

The coordinates of the points of intersection of the lines (4.23) and (4.24) are given by (x, y) :

$$\frac{x}{b_1c_2 - b_2c_1} = \frac{y}{c_1a_2 - c_2a_1} = \frac{1}{a_1b_2 - a_2b_1}$$

$$\therefore \frac{x}{\sin\frac{\alpha+\beta}{2} \cdot 0 - \left(-\cos\frac{\alpha+\beta}{2}\right) \cdot \left(-a\cos\frac{\alpha-\beta}{2}\right)}$$

$$= \frac{y}{\left(-a\cos\frac{\alpha-\beta}{2}\right) \cdot \sin\frac{\alpha+\beta}{2} - 0 \cdot \cos\frac{\alpha+\beta}{2}}$$

$$= \frac{1}{\cos\frac{\alpha+\beta}{2} \cdot \left(-\cos\frac{\alpha+\beta}{2}\right) - \sin\frac{\alpha+\beta}{2} \cdot \sin\frac{\alpha+\beta}{2}}$$

4.4. Bisectors of angles between straight lines

$$\therefore \frac{x}{\cos\dfrac{\alpha+\beta}{2}\cos\dfrac{\alpha-\beta}{2}} = \frac{y}{\sin\dfrac{\alpha+\beta}{2}\cos\dfrac{\alpha-\beta}{2}}$$

$$= \frac{a}{\sin^2\dfrac{\alpha+\beta}{2}+\cos^2\dfrac{\alpha+\beta}{2}} = a$$

$$\therefore x = \frac{a}{2}(\cos\alpha+\cos\beta),\ y = \frac{a}{2}(\sin\alpha+\sin\beta) \quad (4.25)$$

It is easy to see that this is also the middle point of the line joining the given points. ∎

§ Problem 4.4.8. *Find the equations of the two straight lines drawn through the point $(0, a)$ on which the perpendiculars let fall from the point $(2a, 2a)$ are each of length a.*

Prove also that the equation of the straight line joining the feet of these perpendiculars is $y + 2x = 5a$. ◊

§§ Solution. The equation of the line passing through the point $(0, a)$ is

$$y - a = m(x - 0)$$
$$\therefore mx - y + a = 0 \quad (4.26)$$

The length of the perpendicular on it from the point $(2a, 2a)$ is a, i.e.

$$\frac{m\cdot 2a - 2a + a}{\sqrt{m^2+1}} = a$$
$$\therefore 2m - 1 = \sqrt{m^2+1}$$
$$\therefore 4m^2 + 1 - 4m = m^2 + 1$$
$$\therefore 3m^2 - 4m = 0$$
$$\therefore m(3m - 4) = 0$$
$$\therefore m = 0,\ \text{or}\ \frac{4}{3}.$$

Putting these values of m in (4.26), the required equations are

$$y = a,\ \text{and} \quad (4.27)$$
$$\frac{4}{3}x - y + a = 0,\ \text{i.e.}\ 3y = 4x + 3a. \quad (4.28)$$

The line passing through the point $(2a, 2a)$ and perpendicular to (4.27) is

$$x = 2a \quad (4.29)$$

The line passing through the point $(2a, 2a)$ and perpendicular to (4.28) is

$$y - 2a = -\frac{3}{4}(x - 2a)$$
$$\therefore 4y + 3x = 14a. \quad (4.30)$$

From (4.27) and (4.29), the feet of the first perpendicular is $(2a, a)$.

To find the feet of the second perpendicular, let us solve (4.28) and (4.30):

$(4.30) \times 3 - (4.28) \times 4 \implies$

$9x + 16x = 42a - 12a,\ \therefore x = \dfrac{6a}{5}$ and $y = \dfrac{13a}{5}$.

Hence the feet of the perpendiculars are

$$(2a, a)\ \text{and}\ \left(\frac{6a}{5}, \frac{13a}{5}\right).$$

The equation of the line joining these two points are
$$y - a = \frac{\frac{13a}{5} - a}{\frac{6a}{5} - 2a}(x - 2a)$$
$$\therefore y + 2x = 5a. \qquad \blacksquare$$

§ Problem 4.4.9. *Find the point of intersection and the inclination of the two lines*
$$Ax + By = A + B \text{ and } A(x - y) + B(x + y) = 2B. \qquad \diamond$$

§§ Solution. The given equations are
$$Ax + By = A + B, \text{ or } Ax + By - (A + B) = 0, \text{ and} \qquad (4.31)$$
$$A(x - y) + B(x + y) = 2B, \text{ or } (A + B)x + (B - A)y - 2B = 0. \qquad (4.32)$$

The coordinates of the points of intersection are given by (x, y):
$$\frac{x}{b_1 c_2 - b_2 c_1} = \frac{y}{c_1 a_2 - c_2 a_1} = \frac{1}{a_1 b_2 - a_2 b_1}$$

$$\therefore \frac{x}{B \cdot (-2B) - (B - A) \cdot (-(A + B))} = \frac{y}{-(A + B) \cdot (A + B) - (-2B) \cdot A}$$
$$= \frac{1}{A(B - A) - (A + B)B}$$

$$\therefore \frac{x}{-(A^2 + B^2)} = \frac{y}{-(A^2 + B^2)} = \frac{1}{-(A^2 + B^2)}$$
$$\therefore x = 1, \ y = 1.$$

Hence the point of intersection is $(1, 1)$.

Slope of the line (4.31) $= m_1 = -\dfrac{A}{B}$.

Slope of the line (4.32) $= m_2 = -\dfrac{A + B}{B - A}$.

Hence the angle between these two lines
$$= \tan^{-1} \frac{m_1 - m_2}{1 + m_1 m_2} = \tan^{-1} \frac{-\dfrac{A}{B} + \dfrac{A + B}{B - A}}{1 + \dfrac{A}{B} \cdot \dfrac{A + B}{B - A}}$$
$$= \tan^{-1} \frac{-AB + A^2 + AB + B^2}{B^2 - AB + A^2 + AB} = \tan^{-1} 1 = 45°. \qquad \blacksquare$$

§ Problem 4.4.10. *Find the coordinates of the point in which the line*
$$2y - 3x + 7 = 0$$
meets the line joining the two points $(6, -2)$ and $(-8, 7)$. Find also the angle between them. $\qquad \diamond$

§§ Solution. The equation of the line joining the two points $(6, -2)$ and $(-8, 7)$ is
$$y - (-2) = \frac{7 - (-2)}{-8 - 6}(x - 6)$$
$$\therefore 9x + 14y - 26 = 0. \qquad (4.33)$$

The equation of the given line is
$$2y - 3x + 7 = 0. \qquad (4.34)$$

$(4.33) - (4.34) \times 7 \implies 30x = 75, \therefore x = \dfrac{5}{2}$, and $y = \dfrac{1}{4}$.

Hence the point is $\left(\dfrac{5}{2}, \dfrac{1}{4}\right)$.

4.4. Bisectors of angles between straight lines

Slope of (4.33) is $m_2 = -\dfrac{9}{14}$.

Slope of (4.34) is $m_1 = \dfrac{3}{2}$.

Hence the angle between these two lines

$$= \tan^{-1} \frac{m_1 - m_2}{1 + m_1 m_2} = \tan^{-1} \frac{\dfrac{3}{2} + \dfrac{9}{14}}{1 - \dfrac{3}{2} \cdot \dfrac{9}{14}} = \tan^{-1} 60.$$
∎

§ Problem 4.4.11. *Find the coordinates of the feet of the perpendiculars let fall from the point $(5,0)$ upon the sides of the triangle formed by joining the three points $(4,3)$, $(-4,3)$ and $(0,-5)$; prove also that the points so determined lie on a straight line.* ◊

§§ Solution. Let the angular points of the given triangle be labeled as $A(4,3)$, $B(-4,3)$, $C(0,-5)$.

The equation of the side AB is

$$y - 3 = \frac{3-3}{4+4}(x-4), \text{ or}, y = 3. \tag{4.35}$$

Its slope $= m_1 = 0$.

The equation of the side BC is

$$y - 3 = \frac{-5-3}{0+4}(x+4), \text{ or}, 2x + y + 5 = 0. \tag{4.36}$$

Its slope $= m_2 = -2$.

The equation of the side CA is

$$y + 5 = \frac{3+5}{4-0}(x-0), \text{ or}, -2x + y + 5 = 0. \tag{4.37}$$

Its slope $= m_3 = 2$.

The equation of the line passing through through the point $(5,0)$ and perpendicular to (4.35) is

$$y - 0 = -\frac{1}{m_1}(x-5), \text{ or } x = 5. \tag{4.38}$$

The equation of the line passing through through the point $(5,0)$ and perpendicular to (4.36) is

$$y - 0 = -\frac{1}{m_2}(x-5), \text{ or } x - 2y = 5. \tag{4.39}$$

The equation of the line passing through through the point $(5,0)$ and perpendicular to (4.37) is

$$y - 0 = -\frac{1}{m_3}(x-5), \text{ or } x + 2y = 5. \tag{4.40}$$

Solving (4.35) and (4.38), the feet of the first perpendicular is $P(5,3)$.

Solving (4.36) and (4.39), the feet of the second perpendicular is $Q(-1,-3)$.

Solving (4.37) and (4.40), the feet of the third perpendicular is $R(3,1)$.

Area of the $\Delta PQR = \dfrac{1}{2}\{5(-3-1) - 1(1-3) + 3(3+3)\} = 0$.

Hence the points P, Q, R are collinear.

Alternatively, we can also find the equation of the line PQ, and prove that the point R satisfies it. ∎

4.4. Bisectors of angles between straight lines

§ Problem 4.4.12. *Find the coordinates of the point of intersection of the straight lines*
$$2x - 3y = 1 \text{ and } 5y - x = 3,$$
and determine also the angle at which they cut one another. ◊

§§ Solution. The given equations are
$$2x - 3y = 1 \text{ and} \quad (4.41)$$
$$5y - x = 3 \quad (4.42)$$
$(4.41) + 2 \times (4.42) \implies y = 1, x = 2.$
Hence the point of intersection is $(2, 1)$.

The slope of (4.41) is $= m_1 = \dfrac{2}{3}$.

The slope of (4.42) is $= m_2 = \dfrac{1}{5}$.

Hence the angle between these two lines
$$= \tan^{-1} \frac{m_1 - m_2}{1 + m_1 m_2} = \tan^{-1} \frac{\dfrac{2}{3} - \dfrac{1}{5}}{1 + \dfrac{2}{3} \cdot \dfrac{1}{5}} = \tan^{-1} \frac{7}{17}.$$
∎

§ Problem 4.4.13. *Find the angle between the two lines*
$$3x + y + 12 = 0 \text{ and } x + 2y - 1 = 0.$$
Find also the coordinates of their point of intersection and the equations of lines drawn perpendicular to them from the point $(3, -2)$. ◊

§§ Solution. The given equations are
$$3x + y + 12 = 0 \text{ and} \quad (4.43)$$
$$x + 2y - 1 = 0 \quad (4.44)$$
The slope of (4.43) is $= m_2 = -3$.

The slope of (4.44) is $= m_1 = -\dfrac{1}{2}$.

Hence the angle between these two lines
$$= \tan^{-1} \frac{m_1 - m_2}{1 + m_1 m_2} = \tan^{-1} \frac{-\dfrac{1}{2} + 3}{1 + 3 \cdot \dfrac{1}{2}} = \tan^{-1} 1 = 45°.$$

$(4.43) - 3 \times (4.44) \implies x = -5, y = 3.$
Hence the point of intersection is $(-5, 3)$.

The equation of the line passing through through the point $(3, -2)$ and perpendicular to (4.43) is
$$y + 2 = -\frac{1}{m_2}(x - 3)$$
$$\therefore y + 2 = \frac{1}{3}(x - 3)$$
$$\therefore x - 3y = 9.$$

The equation of the line passing through through the point $(3, -2)$ and perpendicular to (4.44) is
$$y + 2 = -\frac{1}{m_1}(x - 3)$$
$$\therefore y + 2 = 2(x - 3)$$
$$\therefore 2x - y = 8.$$
∎

§ Problem 4.4.14. *Prove that the points whose coordinates are respectively* $(5, 1)$, $(1, -1)$, *and* $(11, 4)$ *lie on a straight line, and find its intercepts on the axes.* ◊

4.4. Bisectors of angles between straight lines

§§ Solution. Let the points be labeled as $A(5,1)$, $B(1,-1)$, $C(11,4)$.
Area of the $\triangle ABC = \dfrac{1}{2}\{5(-1-4) + 1(4-1) + 11(1+1)\} = 0$.
Hence the points A, B, C are collinear.
The equation of the line AB is
$$y - 1 = \frac{-1-1}{1-5}(x-5) = \frac{1}{2}(x-5)$$
$$\therefore x - 2y = 3.$$
Putting $x = 0$, the y-intercept $= -\dfrac{3}{2}$.
Putting $y = 0$, the x-intercept $= 3$.
Alternatively, it is easy to see that the line $AB : x - 2y = 3$ passes through $C(11, 4)$. ∎

Prove that the following sets of three lines meet in a point.

§ Problem 4.4.15. $2x - 3y = 7$, $3x - 4y = 13$, and $8x - 11y = 33$. ◊

§§ Solution. By $Art.$ 79, let us compute the following
$$\begin{vmatrix} a_1 & b_1 & c_1 \\ a_2 & b_2 & c_2 \\ a_3 & b_3 & c_3 \end{vmatrix} = \begin{vmatrix} 2 & -3 & -7 \\ 3 & -4 & -13 \\ 8 & -11 & -33 \end{vmatrix}$$
$= 2(4 \times 33 - 11 \times 13) + 3(3 \times -33 + 8 \times 13) - 7(3 \times -11 + 8 \times 4)$
$= 2(132 - 143) + 3(-99 + 104) - 7(-33 + 32)$
$= 2(-11) + 3(5) - 7(-1)$
$= -22 + 15 + 7$
$= 0$.
∴ the lines are concurrent. [$Art.$ 79.]

Alternative Solution :
It is easy to see that
$(2x - 3y - 7) + 2(3x - 4y - 13) - (8x - 11y - 33) = 0$.
∴ the lines are concurrent. [$Art.$ 80.]

Alternative Solution :
The given lines are
$$2x - 3y = 7 \tag{4.45}$$
$$3x - 4y = 13 \tag{4.46}$$
$$8x - 11y = 33 \tag{4.47}$$
(4.45) × 3 − (4.46) × 2 $\implies$ $y = 5$, $x = 11$.
Hence the point of intersection of (4.45) and (4.46) is $(11, 5)$.
It is easy to see that this point lies on the line (4.47). Hence the three lines meet in a point. ∎

§ Problem 4.4.16. $3x + 4y + 6 = 0$, $6x + 5y + 9 = 0$, and $3x + 3y + 5 = 0$. ◊

§§ Solution. By $Art.$ 79, let us compute the following
$$\begin{vmatrix} a_1 & b_1 & c_1 \\ a_2 & b_2 & c_2 \\ a_3 & b_3 & c_3 \end{vmatrix} = \begin{vmatrix} 3 & 4 & 6 \\ 6 & 5 & 9 \\ 3 & 3 & 5 \end{vmatrix}$$
$= 3(5 \times 5 - 9 \times 3) - 4(6 \times 5 - 9 \times 3) + 6(6 \times 3 - 5 \times 3)$
$= 0$.
∴ the lines are concurrent. [$Art.$ 79.]

4.4. Bisectors of angles between straight lines

Alternative Solution :
It is easy to see that
$$(3x + 4y + 6) + (6x + 5y + 9) - 3(3x + 3y + 5) = 0.$$
∴ the lines are concurrent. [Art. 80.] ∎

§ Problem 4.4.17. $\frac{x}{a} + \frac{y}{b} = 1$, $\frac{x}{b} + \frac{y}{a} = 1$, and $y = x$. ◊

§§ Solution. By *Art.* 79, let us compute the following
$$\begin{vmatrix} a_1 & b_1 & c_1 \\ a_2 & b_2 & c_2 \\ a_3 & b_3 & c_3 \end{vmatrix} = \begin{vmatrix} \frac{1}{a} & \frac{1}{b} & -1 \\ \frac{1}{b} & \frac{1}{a} & -1 \\ 1 & -1 & 0 \end{vmatrix} = 0.$$
∴ the lines are concurrent. [*Art.* 79.]

Alternative Solution :
It is easy to see that
$$ab\left\{\frac{x}{a} + \frac{y}{b} - 1\right\} - ab\left\{\frac{x}{b} + \frac{y}{a} - 1\right\} + (a-b)\{x - y\} = 0.$$
∴ the lines are concurrent. [*Art.* 80.] ∎

§ Problem 4.4.18. *Prove that the three straight lines whose equations are*
$$15x - 18y + 1 = 0, \ 12x + 10y - 3 = 0, \ and \ 6x + 66y - 11 = 0$$
all meet in a point.

Show also that the third line bisects the angle between the other two. ◊

§§ Solution. By *Art.* 79, let us compute the following
$$\begin{vmatrix} a_1 & b_1 & c_1 \\ a_2 & b_2 & c_2 \\ a_3 & b_3 & c_3 \end{vmatrix} = \begin{vmatrix} 15 & -18 & 1 \\ 12 & 10 & -3 \\ 6 & 66 & -11 \end{vmatrix} = 0.$$
∴ the lines are concurrent. [*Art.* 79.]

The bisectors of the angle between the first two lines are
$$\frac{15x - 18y + 1}{\sqrt{15^2 + 18^2}} = \pm\frac{12x + 10y - 3}{\sqrt{12^2 + 10^2}}$$
$$\therefore \frac{15x - 18y + 1}{3\sqrt{61}} = \pm\frac{12x + 10y - 3}{2\sqrt{61}}$$

Taking the upper sign, we have $6x + 66y - 11 = 0$, which is the same as the equation of the third line. Hence the third line bisects the angle between the other two lines.

Alternative Solution :
It is easy to see that
$$2(15x - 18y + 1) - 3(12x + 10y - 3) + (6x + 66y - 11) = 0.$$
∴ the lines are concurrent. [*Art.* 80.] ∎

§ Problem 4.4.19. *Find the conditions that the straight lines*
$$y = m_1 x + a_1, \ y = m_2 x + a_2, \ and \ y = m_3 x + a_3$$
may meet in a point. ◊

§§ Solution. By *Art.* 79, let us compute the following
$$\begin{vmatrix} a_1 & b_1 & c_1 \\ a_2 & b_2 & c_2 \\ a_3 & b_3 & c_3 \end{vmatrix} = \begin{vmatrix} m_1 & -1 & a_1 \\ m_2 & -1 & a_2 \\ m_3 & -1 & a_3 \end{vmatrix}$$
$$= m_1(-a_3 + a_2) + (m_2 a_3 - m_3 a_2) + a_1(-m_2 + m_3)$$
$$= m_1(a_2 - a_3) + m_2(a_3 - a_1) + m_3(a_1 - a_2)$$

4.4. Bisectors of angles between straight lines

Hence the condition of the collinearity is the following
$$m_1(a_2 - a_3) + m_2(a_3 - a_1) + m_3(a_1 - a_2) = 0.$$ ∎

Find the coordinates of the orthocentre of the triangles whose angular points are

§ Problem 4.4.20. $(0,0)$, $(2,-1)$ and $(-1,3)$. ◊

§§ Solution. Let the angular points of the triangle be $A(0,0)$, $B(2,-1)$, $C(-1,3)$.

The equation of the side AB is
$$y - 0 = \frac{-1-0}{2-0}(x-0)$$
$$\therefore x + 2y = 0 \tag{4.48}$$

Its slope $= -\frac{1}{2}$.

Equation of the line perpendicular on the side AB: (4.48) from $C(-1,3)$ is
$$y - 3 = 2(x+1)$$
$$\therefore 2x - y + 5 = 0 \tag{4.49}$$

The equation of the side AC is
$$y - 0 = \frac{3-0}{-1-0}(x-0)$$
$$\therefore 3x + y = 0 \tag{4.50}$$

Its slope $= -3$.

Equation of the line perpendicular on the side AC: (4.50) from $B(2,-1)$ is
$$y + 1 = \frac{1}{3}(x-2)$$
$$\therefore x - 3y - 5 = 0 \tag{4.51}$$

(4.49) $- 2 \times$ (4.51) $\implies$
$$-y + 6y + 5 + 10 = 0$$
$$\therefore y = -3$$
$$\therefore x = 5 + 3y = 5 - 9 = -4$$

It is clear to see that the point of intersection of the lines (4.49) and (4.51) is the orthocentre of the triangle ABC, the coordinates of the orthocentre are : $(-4,-3)$. ∎

§ Problem 4.4.21. $(0,0)$, $(2,-1)$ and $(-1,3)$. ◊

§§ Solution. Let the angular points of the triangle be $A(1,0)$, $B(2,-4)$, $C(-5,-2)$.

The equation of the side AB is
$$y - 0 = \frac{-4-0}{2-1}(x-1)$$
$$\therefore 4x + y - 4 = 0 \tag{4.52}$$

Its slope $= -4$.

Equation of the line perpendicular on the side AB: (4.52) from $C(-5,-2)$ is
$$y + 2 = \frac{1}{4}(x+5)$$
$$\therefore -x + 4y + 3 = 0 \tag{4.53}$$

4.4. Bisectors of angles between straight lines

The equation of the side AC is
$$y - 0 = \frac{-2-0}{-5-1}(x-1)$$
$$\therefore x - 3y - 1 = 0 \tag{4.54}$$

Its slope $= \frac{1}{3}$.

Equation of the line perpendicular on the side AC: (4.54) from $B(2, -4)$ is
$$y + 4 = -3(x - 2)$$
$$\therefore 3x + y - 2 = 0 \tag{4.55}$$

$(4.53) \times 3 + (4.55) \implies$
$$-3x + 12y + 9 + 3x + y - 2 = 0$$
$$\therefore 13y + 7 = 0$$
$$\therefore y = -\frac{7}{13}$$
$$\therefore x = 4y + 3 = 3 - \frac{28}{13} = \frac{11}{13}$$

It is clear to see that the point of intersection of the lines (4.53) and (4.55) is the orthocentre of the triangle ABC, the coordinates of the orthocentre are : $\left(\frac{11}{13}, -\frac{7}{13}\right)$. ∎

§ Problem 4.4.22. *In any triangle ABC, prove that*

(1) the bisectors of the angles A, B, and C meet in a point,

(2) the medians, i.e. the lines joining each vertex to the middle point of the opposite side, meet in a point, and

(3) the straight lines through the middle points of the sides perpendicular to the sides meet in a point. ◊

§§ Solution. (1) Let the equations of the sides of the triangle be
$$x \cos \alpha_1 + y \sin \alpha_1 = p_1$$
$$x \cos \alpha_2 + y \sin \alpha_2 = p_2$$
$$x \cos \alpha_3 + y \sin \alpha_3 = p_3$$

Let us simplify these equations by substituting the following:
$$c_1 = \cos \alpha_1, \ c_2 = \cos \alpha_2, \ c_3 = \cos \alpha_3,$$
$$s_1 = \sin \alpha_1, \ s_2 = \sin \alpha_2, \ s_3 = \sin \alpha_3$$

The equations of the sides become:
$$xc_1 + ys_1 = p_1$$
$$xc_2 + ys_2 = p_2$$
$$xc_3 + ys_3 = p_3$$

By *Art.* 84, the equations of the bisectors of the angles are:
$$x(c_1 \pm c_2) + y(s_1 \pm s_2) = p_1 \pm p_2 \tag{4.56}$$
$$x(c_2 \pm c_3) + y(s_2 \pm s_3) = p_2 \pm p_3 \tag{4.57}$$
$$x(c_3 \pm c_1) + y(s_3 \pm s_1) = p_3 \pm p_1 \tag{4.58}$$

Case 1 : If the origin lies within the triangle, then using the lower sign in each case, the above equations become
$$x(c_1 - c_2) + y(s_1 - s_2) = p_1 - p_2 \tag{4.59}$$
$$x(c_2 - c_3) + y(s_2 - s_3) = p_2 - p_3 \tag{4.60}$$
$$x(c_3 - c_1) + y(s_3 - s_1) = p_3 - p_1 \tag{4.61}$$

4.4. Bisectors of angles between straight lines

Adding these equations, it is easy to see that the sum of the coefficients of x, y as well as the sum of the constants become identically zero, hence these lines are concurrent. [$Art.$ 80.]

Case 2 : If the origin lies outside the triangle, then using the upper sign in two cases, say (4.56) and (4.57) and the lower sign in the third case, say (4.58), these equations become

$$x(c_1 + c_2) + y(s_1 + s_2) = p_1 + p_2 \tag{4.62}$$
$$x(c_2 + c_3) + y(s_2 + s_3) = p_2 + p_3 \tag{4.63}$$
$$x(c_3 - c_1) + y(s_3 - s_1) = p_3 - p_1 \tag{4.64}$$

Now (4.62) − (4.63) + (4.64) yields the sum of the coefficients of x, y as well as the sum of the constants become identically zero, hence these lines are concurrent. [$Art.$ 80.]

Alternative Solution : Let the equations of the sides of the triangle be

$$a_1 x + b_1 y + c_1 = 0 \tag{4.65}$$
$$a_2 x + b_2 y + c_2 = 0 \tag{4.66}$$
$$a_3 x + b_3 y + c_3 = 0 \tag{4.67}$$

The equation of the bisectors of (4.65) and (4.66) are

$$\frac{a_1 x + b_1 y + c_1}{\sqrt{a_1^2 + b_1^2}} = \pm \frac{a_2 x + b_2 y + c_2}{\sqrt{a_2^2 + b_2^2}}$$

$$\therefore x \left(\frac{a_1}{\sqrt{a_1^2 + b_1^2}} \mp \frac{a_2}{\sqrt{a_2^2 + b_2^2}} \right) + y \left(\frac{b_1}{\sqrt{a_1^2 + b_1^2}} \mp \frac{b_2}{\sqrt{a_2^2 + b_2^2}} \right)$$
$$+ \left(\frac{c_1}{\sqrt{a_1^2 + b_1^2}} \mp \frac{c_2}{\sqrt{a_2^2 + b_2^2}} \right) = 0$$

Similarly, the equations of other bisectors are

$$\therefore x \left(\frac{a_2}{\sqrt{a_2^2 + b_2^2}} \mp \frac{a_3}{\sqrt{a_3^2 + b_3^2}} \right) + y \left(\frac{b_2}{\sqrt{a_2^2 + b_2^2}} \mp \frac{b_3}{\sqrt{a_3^2 + b_3^2}} \right)$$
$$+ \left(\frac{c_2}{\sqrt{a_2^2 + b_2^2}} \mp \frac{c_3}{\sqrt{a_3^2 + b_3^2}} \right) = 0$$

$$\therefore x \left(\frac{a_3}{\sqrt{a_3^2 + b_3^2}} \mp \frac{a_1}{\sqrt{a_1^2 + b_1^2}} \right) + y \left(\frac{b_3}{\sqrt{a_3^2 + b_3^2}} \mp \frac{b_1}{\sqrt{a_1^2 + b_1^2}} \right)$$
$$+ \left(\frac{c_3}{\sqrt{a_3^2 + b_3^2}} \mp \frac{c_1}{\sqrt{a_1^2 + b_1^2}} \right) = 0$$

Let us simplify these equations by substituting the following:

$$\alpha = \frac{1}{\sqrt{a_1^2 + b_1^2}}, \ \beta = \frac{1}{\sqrt{a_2^2 + b_2^2}}, \ \gamma = \frac{1}{\sqrt{a_3^2 + b_3^2}},$$

The equations of the sides become :

$$x(a_1 \alpha \mp a_2 \beta) + y(b_1 \alpha \mp b_2 \beta) + (c_1 \alpha \mp c_2 \beta) = 0 \tag{4.68}$$
$$x(a_2 \beta \mp a_3 \gamma) + y(b_2 \beta \mp b_3 \gamma) + (c_2 \beta \mp c_3 \gamma) = 0 \tag{4.69}$$
$$x(a_3 \gamma \mp a_1 \alpha) + y(b_3 \gamma \mp b_1 \alpha) + (c_3 \gamma \mp c_1 \alpha) = 0 \tag{4.70}$$

Case 1 : If the origin lies within the triangle, then using the negative sign in each case, the above equations become

$$x(a_1 \alpha - a_2 \beta) + y(b_1 \alpha - b_2 \beta) + (c_1 \alpha - c_2 \beta) = 0 \tag{4.71}$$

4.4. Bisectors of angles between straight lines

$$x(a_2\beta - a_3\gamma) + y(b_2\beta - b_3\gamma) + (c_2\beta - c_3\gamma) = 0 \qquad (4.72)$$
$$x(a_3\gamma - a_1\alpha) + y(b_3\gamma - b_1\alpha) + (c_3\gamma - c_1\alpha) = 0 \qquad (4.73)$$

Adding these equations, it is easy to see that the sum of the coefficients of x, y as well as the sum of the constants become identically zero, hence these lines are concurrent. [$Art.$ 80.]

Case 2 : If the origin lies outside the triangle, then using the positive sign in two cases, say (4.68) and (4.69) and the lower sign in the third case, say (4.70), these equations become

$$x(a_1\alpha + a_2\beta) + y(b_1\alpha + b_2\beta) + (c_1\alpha + c_2\beta) = 0 \qquad (4.74)$$
$$x(a_2\beta + a_3\gamma) + y(b_2\beta + b_3\gamma) + (c_2\beta + c_3\gamma) = 0 \qquad (4.75)$$
$$x(a_3\gamma - a_1\alpha) + y(b_3\gamma - b_1\alpha) + (c_3\gamma - c_1\alpha) = 0 \qquad (4.76)$$

Now (4.74) − (4.75) + (4.76) yields the sum of the coefficients of x, y as well as the sum of the constants become identically zero, hence these lines are concurrent. [$Art.$ 80.]

(2) Let the coordinates of the angular points of the triangle be $(2x_1, 2y_1)$, $(2x_2, 2y_2)$, $(2x_3, 2y_3)$.

Then the coordinates of the middle points of the sides are

$(x_1 + x_2, y_1 + y_2)$, $(x_2 + x_3, y_2 + y_3)$, $(x_3 + x_1, y_3 + y_1)$.

The equation of the median passing through the points $(2x_1, 2y_1)$ and $(x_2 + x_3, y_2 + y_3)$ is

$$y - 2y_1 = \frac{y_2 + y_3 - 2y_1}{x_2 + x_3 - 2x_1}(x - 2x_1)$$

$$x(y_2 + y_3 - 2y_1) - y(x_2 + x_3 - 2x_1) - 2x_1(y_2 + y_3) + 2y_1(x_2 + x_3) = 0 \qquad (4.77)$$

Similarly the equations of the other two medians are

$$x(y_3 + y_1 - 2y_2) - y(x_3 + x_1 - 2x_2) - 2x_2(y_3 + y_1) + 2y_2(x_3 + x_1) = 0 \qquad (4.78)$$
$$x(y_1 + y_2 - 2y_3) - y(x_1 + x_2 - 2x_3) - 2x_3(y_1 + y_2) + 2y_3(x_1 + x_2) = 0 \qquad (4.79)$$

Adding these equations, it is easy to see that the sum of the coefficients of x, y as well as the sum of the constants become identically zero, hence these three lines are concurrent. [$Art.$ 80.]

(3) Let the coordinates of the angular points of the triangle be $(2x_1, 2y_1)$, $(2x_2, 2y_2)$, $(2x_3, 2y_3)$.

Then the coordinates of the middle points of the sides are

$(x_1 + x_2, y_1 + y_2)$, $(x_2 + x_3, y_2 + y_3)$, $(x_3 + x_1, y_3 + y_1)$.

Slope of the line joining the points $(2x_1, 2y_1)$ and $(2x_2, 2y_2)$ is $= \dfrac{y_2 - y_1}{x_2 - x_1}$.

Hence the slope of the line perpendicular to the above line is $= -\dfrac{x_2 - x_1}{y_2 - y_1}$.

Hence the equation of the perpendicular to the line joining $(2x_1, 2y_1)$ and $(2x_2, 2y_2)$ through $(x_1 + x_2, y_1 + y_2)$, is

$$y - (y_1 + y_2) = -\frac{x_2 - x_1}{y_2 - y_1}(x - x_1 - x_2)$$

$$\therefore x(x_1 - x_2) + y(y_1 - y_2) - (x_1^2 - x_2^2) + (y_1^2 - y_2^2) = 0 \qquad (4.80)$$

4.4. Bisectors of angles between straight lines

Similarly the other two perpendiculars are
$$x(x_2 - x_3) + y(y_2 - y_3) - (x_2^2 - x_3^2) + (y_2^2 - y_3^2) = 0 \qquad (4.81)$$
$$x(x_3 - x_1) + y(y_3 - y_1) - (x_3^2 - x_1^2) + (y_3^2 - y_1^2) = 0 \qquad (4.82)$$

Adding these equations, it is easy to see that the sum of the coefficients of x, y as well as the sum of the constants become identically zero, hence these three lines are concurrent. [*Art.* 80.] ∎

Find the equation to the straight line passing through

§ Problem 4.4.23. *the point* $(3, 2)$ *and the point of intersection of the lines*
$$2x + 3y = 1 \text{ and } 3x - 4y = 6. \qquad \diamond$$

§§ Solution. By *Art.* 82., the equation to the line passing through the point of intersection of the given lines is
$$(2x + 3y - 1) + \lambda(3x - 4y - 6) = 0$$

Since this line passes through the given point $(3, 2)$:
$$\therefore (6 + 6 - 1) + \lambda(9 - 8 - 6) = 0$$
$$\therefore \lambda = \frac{11}{5}.$$

Hence the equation becomes
$$5(2x + 3y - 1) + 11(3x - 4y - 6) = 0$$
$$\therefore 43x - 29y = 71. \qquad \blacksquare$$

§ Problem 4.4.24. *the point* $(2, -9)$ *and the intersection of the lines*
$$2x + 5y - 8 = 0 \text{ and } 3x - 4y = 35. \qquad \diamond$$

§§ Solution. By *Art.* 82., the equation to the line passing through the point of intersection of the given lines is
$$(2x + 5y - 8) + \lambda(3x - 4y - 35) = 0$$

Since this line passes through the given point $(2, -9)$:
$$\therefore (4 - 45 - 8) + \lambda(6 + 36 - 35) = 0$$
$$\therefore \lambda = \frac{49}{7} = 7.$$

Hence the equation becomes
$$(2x + 5y - 8) + 7(3x - 4y - 35) = 0$$
$$\therefore 23x - 23y = 253$$
$$\therefore x - y = 11. \qquad \blacksquare$$

§ Problem 4.4.25. *the origin and the point of intersection of*
$$x - y - 4 = 0 \text{ and } 7x + y + 20 = 0,$$
proving that it bisects the angle between them. $\diamond$

§§ Solution. Solving the given equations, we get the point of intersection as $(-2, -6)$.

Hence the equation of the line passing through this point and the origin is
$$y - 0 = \frac{-6 - 0}{-2 - 0}(x - 0)$$
$$\therefore y = 3x.$$

The equation of one of the bisectors of the given lines in the question is
$$\frac{x - y - 4}{\sqrt{1 + 1}} = -\frac{7x + y + 20}{\sqrt{49 + 1}}$$
$$\therefore 5(x - y - 4) = -(7x + y + 20)$$

4.4. Bisectors of angles between straight lines

$$\therefore 12x = 4y$$
$$\therefore y = 3x. \qquad \blacksquare$$

§ Problem 4.4.26. *the origin and the point of intersection of the lines*
$$\frac{x}{a} + \frac{y}{b} = 1 \text{ and } \frac{x}{b} + \frac{y}{a} = 1. \qquad \diamond$$

§§ Solution. By *Art.* 82., the equation to the line passing through the point of intersection of the given lines is
$$\left(\frac{x}{a} + \frac{y}{b} - 1\right) + \lambda\left(\frac{x}{b} + \frac{y}{a} - 1\right) = 0$$
Since it passes through the origin $(0,0)$,
$$(0 + 0 - 1) + \lambda(0 + 0 - 1) = 0$$
$$\therefore \lambda = -1.$$
Hence the equation becomes
$$\left(\frac{x}{a} + \frac{y}{b} - 1\right) - \left(\frac{x}{b} + \frac{y}{a} - 1\right) = 0$$
$$\therefore x\left(\frac{1}{a} - \frac{1}{b}\right) = y\left(\frac{1}{a} - \frac{1}{b}\right)$$
$$\therefore y = x. \qquad \blacksquare$$

§ Problem 4.4.27. *the point (a, b) and the intersection of the same two lines.* $\qquad \diamond$

§§ Solution. By *Art.* 82., the equation to the line passing through the point of intersection of the given lines is
$$\left(\frac{x}{a} + \frac{y}{b} - 1\right) + \lambda\left(\frac{x}{b} + \frac{y}{a} - 1\right) = 0$$
Since it passes through the origin (a, b),
$$\left(\frac{a}{a} + \frac{b}{b} - 1\right) + \lambda\left(\frac{a}{b} + \frac{b}{a} - 1\right) = 0$$
$$\therefore 1 + \lambda\left(\frac{a^2 + b^2}{ab} - 1\right) = 0.$$
$$\therefore \lambda = \frac{ab}{ab - a^2 - b^2}.$$
Hence the equation becomes
$$\left(\frac{x}{a} + \frac{y}{b} - 1\right) + \frac{ab}{ab - a^2 - b^2}\left(\frac{x}{b} + \frac{y}{a} - 1\right) = 0$$
$$\therefore x\left(\frac{1}{a} + \frac{a}{ab - a^2 - b^2}\right) + y\left(\frac{1}{b} + \frac{b}{ab - a^2 - b^2}\right) - \left(\frac{2ab - a^2 - b^2}{ab - a^2 - b^2}\right) = 0$$
$$\therefore x\left(\frac{ab - b^2}{a(ab - a^2 - b^2)}\right) + y\left(\frac{ab - a^2}{b(ab - a^2 - b^2)}\right) + \frac{(a-b)^2}{ab - a^2 - b^2} = 0$$
$$\therefore a^2 y - b^2 x = ab(a - b).$$

Alternative Solution : The equations of the lines are
$$\frac{x}{a} + \frac{y}{b} = 1 \qquad (4.83)$$
$$\frac{x}{b} + \frac{y}{a} = 1 \qquad (4.84)$$
$(4.83) \times \dfrac{1}{b} - (4.84) \times \dfrac{1}{a} \implies$
$$y\left(\frac{1}{b^2} - \frac{1}{a^2}\right) = \frac{1}{b} - \frac{1}{a}$$

4.4. Bisectors of angles between straight lines

$$\therefore y = \frac{ab}{a+b}$$
$$\therefore x = a\left(1 - \frac{y}{b}\right) = \frac{ab}{a+b}$$

Hence the point of intersection is $\left(\frac{ab}{a+b}, \frac{ab}{a+b}\right)$.

Hence the required equation passing through the points (a, b) and $\left(\frac{ab}{a+b}, \frac{ab}{a+b}\right)$ is

$$y - b = \frac{\dfrac{ab}{a+b} - b}{\dfrac{ab}{a+b} - a}(x - a)$$

$$\therefore y - b = \frac{-b^2}{-a^2}(x - a)$$

$$\therefore a^2(y - b) = b^2(x - a)$$

$$\therefore a^2 y - b^2 x = ab(a - b). \qquad \blacksquare$$

§ Problem 4.4.28. *the intersection of the lines*
$$x - 2y - a = 0 \text{ and } x + 3y - 2a = 0$$
and parallel to the straight line
$$3x + 4y = 0. \qquad \diamond$$

§§ Solution. By *Art.* 82., the equation to the line passing through the point of intersection of the given lines is
$$(x - 2y - a) + \lambda(x + 3y - 2a) = 0 \qquad (4.85)$$

Its slope $= \dfrac{1 + \lambda}{2 - 3\lambda}$.

Slope of the line $3x + 4y = 0$ is $-\dfrac{3}{4}$.

Since (4.85) is parallel to the line $3x + 4y = 0$, hence
$$\frac{1 + \lambda}{2 - 3\lambda} = -\frac{3}{4}$$
$$\therefore 4 + 4\lambda = -6 + 9\lambda$$
$$\therefore \lambda = 2$$

Hence the required equation (4.85) becomes
$$(x - 2y - a) + 2(x + 3y - 2a) = 0$$
$$\therefore 3x + 4y = 5a. \qquad \blacksquare$$

§ Problem 4.4.29. *the intersection of the lines*
$$x + 2y + 3 = 0 \text{ and } 3x + 4y + 7 = 0$$
and perpendicular to the straight line
$$y - x = 8. \qquad \diamond$$

§§ Solution. By *Art.* 82., the equation to the line passing through the point of intersection of the given lines is
$$(x + 2y + 3) + \lambda(3x + 4y + 7) = 0$$
$$\therefore x(1 + 3\lambda) + y(2 + 4\lambda) + (3 + 7\lambda) = 0$$

Its slope is $= -\dfrac{1 + 3\lambda}{2 + 4\lambda}$.

Since this is perpendicular to $y - x = 8$,
$$1 + 3\lambda = 2 + 4\lambda$$
$$\therefore \lambda = -1.$$

4.4. Bisectors of angles between straight lines

Hence the equation becomes
$$-2x - 2y - 4 = 0$$
$$\therefore x + y + 2 = 0.$$
∎

§ Problem 4.4.30. *the intersection of the lines*
$$3x - 4y + 1 = 0 \text{ and } 5x + y - 1 = 0$$
and cutting off equal intercepts from the axes. ◊
§§ Solution. By *Art.* 82., the equation to the line passing through the point of intersection of the given lines is
$$(3x - 4y + 1) + \lambda(5x + y - 1) = 0$$
$$\therefore x(3 + 5\lambda) + y(-4 + \lambda) = \lambda - 1$$
$$\therefore \frac{x}{\dfrac{\lambda - 1}{3 + 5\lambda}} + \frac{y}{\dfrac{\lambda - 1}{-4 + \lambda}} = 1$$

Since this cuts off equal intercepts from the axes, comparing this with the equation $\dfrac{x}{a} + \dfrac{y}{b} = 1$ it should be $a = b$.
$$\frac{\lambda - 1}{3 + 5\lambda} = \frac{\lambda - 1}{-4 + \lambda}$$
$$\therefore 3 + 5\lambda = -4 + \lambda$$
$$\therefore \lambda = -\frac{7}{4}.$$

Hence the equation becomes
$$(3x - 4y + 1) - \frac{7}{4}(5x + y - 1) = 0$$
$$\therefore 23x + 23y = 11.$$
∎

§ Problem 4.4.31. *the intersection of the lines*
$$2x - 3y = 10 \text{ and } x + 2y = 6$$
and the intersection of the lines
$$16x - 10y = 33 \text{ and } 12x + 14y + 29 = 0.$$
◊
§§ Solution. The first set of the equations are
$$2x - 3y = 10 \qquad (4.86)$$
$$x + 2y = 6 \qquad (4.87)$$
$(4.86) - (4.87) \times 2 \implies$
$$-7y = -2$$
$$\therefore y = \frac{2}{7}$$
$$\therefore x = 6 - 2y = 6 - \frac{4}{7} = \frac{38}{7}$$

Hence the coordinates of the points of intersection of these lines are $\left(\dfrac{38}{7}, \dfrac{2}{7}\right)$.

The first set of the equations are
$$16x - 10y = 33 \qquad (4.88)$$
$$12x + 14y + 29 = 0 \qquad (4.89)$$
$(4.88) \times 3 - (4.89) \times 4 \implies$
$$-86y = 215$$
$$\therefore y = -\frac{5}{2}$$
$$\therefore 16x = 33 + 10y = 33 - \frac{50}{2} = 8$$

4.4. Bisectors of angles between straight lines

$$\therefore x = \frac{1}{2}$$

Hence the coordinates of the points of intersection of these lines are $\left(\frac{1}{2}, -\frac{5}{2}\right)$.

Hence the required equation is

$$y + \frac{5}{2} = \frac{\frac{2}{7} + \frac{5}{2}}{\frac{38}{7} - \frac{1}{2}}\left(x - \frac{1}{2}\right)$$

$$\therefore y + \frac{5}{2} = \frac{4 + 35}{76 - 7}\left(x - \frac{1}{2}\right) = \frac{13}{23}\left(x - \frac{1}{2}\right)$$

$$\therefore 13x - 23y = 64. \quad \blacksquare$$

§ Problem 4.4.32. *If through the angular points of a triangle straight lines be drawn parallel to the sides, and if the intersections of these lines be joined to the opposite angular points of the triangle, show that the joining lines so obtained will meet in a point.* ◊

§§ Solution. Let ABC be the triangle. Let us draw the parallel lines such that

- the parallels passing through B and C meet at a point A'.
- the parallels passing through C and A meet at a point B'.
- the parallels passing through A and B meet at a point C'.

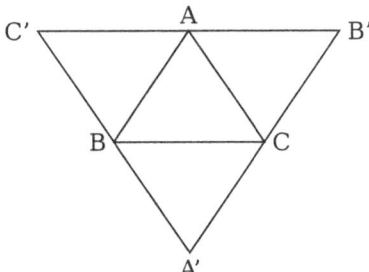

It is obvious to see that $ABCB'$ and $ACBC'$ are parallelograms,
$$\therefore AB' = BC = C'A.$$
$\therefore A, B, C$ are the middle points of the sides $B'C'$, $C'A'$, $A'B'$.

Hence, as already solved in the exercise 22 : (2), the lines AA', BB', and CC' are concurrent.

Let us redo the exercise here for the sake of simplicity.

Let the coordinates of the angular points of the triangle $A'B'C'$ be $(2x_1, 2y_1)$, $(2x_2, 2y_2)$, $(2x_3, 2y_3)$.

Then the coordinates of the middle points of the sides are
$$(x_1 + x_2, y_1 + y_2), (x_2 + x_3, y_2 + y_3), (x_3 + x_1, y_3 + y_1).$$

The equation of the median passing through the points $(2x_1, 2y_1)$ and $(x_2 + x_3, y_2 + y_3)$ is

$$y - 2y_1 = \frac{y_2 + y_3 - 2y_1}{x_2 + x_3 - 2x_1}(x - 2x_1)$$

4.4. Bisectors of angles between straight lines

$$x(y_2 + y_3 - 2y_1) - y(x_2 + x_3 - 2x_1) - 2x_1(y_2 + y_3) + 2y_1(x_2 + x_3) = 0 \tag{4.90}$$

Similarly the equations of the other two medians are
$$x(y_3 + y_1 - 2y_2) - y(x_3 + x_1 - 2x_2) - 2x_2(y_3 + y_1) + 2y_2(x_3 + x_1) = 0 \tag{4.91}$$
$$x(y_1 + y_2 - 2y_3) - y(x_1 + x_2 - 2x_3) - 2x_3(y_1 + y_2) + 2y_3(x_1 + x_2) = 0 \tag{4.92}$$

Adding these equations, it is easy to see that the sum of the coefficients of x, y as well as the sum of the constants become identically zero, hence these three lines are concurrent. [*Art.* 80.] ∎

§ Problem 4.4.33. *Find the equations to the straight lines passing through the point of intersection of the straight lines*
$$Ax + By + C = 0 \text{ and } A'x + B'y + C' = 0 \text{ and}$$

(1) passing through the origin,

(2) parallel to the axis of y,

(3) cutting off a given distance a from the axis of y, and

(4) passing through a given point (x', y'). ◊

§§ Solution. By *Art.* 82., the equation to the line passing through the point of intersection of the given lines is
$$(Ax + By + C) + \lambda(A'x + B'y + C') = 0$$
$$\therefore x(A + \lambda A') + y(B + \lambda B') + (C + \lambda C') = 0$$

(1) If it passes through the origin $(0,0)$, hence $C + \lambda C' = 0$, $\therefore \lambda = -\dfrac{C}{C'}$.

Hence the required equation is
$$(Ax + By + C) - \frac{C}{C'}(A'x + B'y + C') = 0$$

(2) If it is parallel to the axis of y, hence $B + \lambda B' = 0, \therefore \lambda = -\dfrac{B}{B'}$.
Hence the required equation is
$$(Ax + By + C) - \frac{B}{B'}(A'x + B'y + C') = 0$$

(3) Let us rewrite the equation in the format of $\dfrac{x}{a} + \dfrac{y}{b} = 1$, in which the y-intercept is b :
$$x(A + \lambda A') + y(B + \lambda B') + (C + \lambda C') = 0$$
$$\therefore \frac{x}{-\dfrac{C + \lambda C'}{A + \lambda A'}} + \frac{y}{-\dfrac{C + \lambda C'}{B + \lambda B'}} = 1$$

It is clear that the y-intercept is $-\dfrac{C + \lambda C'}{B + \lambda B'}$. It is given as a in the question,
$$-\frac{C + \lambda C'}{B + \lambda B'} = a$$
$$\therefore \lambda = -\frac{Ba + C}{B'a + C'}.$$

4.4. Bisectors of angles between straight lines

Hence the required equation is
$$(Ax + By + C) - \frac{Ba + C}{B'a + C'}(A'x + B'y + C') = 0$$
(4) If it passes through the point (x', y'),
$$\therefore (Ax' + By' + C) + \lambda(A'x' + B'y' + C') = 0$$
$$\therefore \lambda = -\frac{Ax' + By' + C}{A'x' + B'y' + C'}.$$
Hence the required equation is
$$(Ax + By + C) - \frac{Ax' + By' + C}{A'x' + B'y' + C'}(A'x + B'y + C') = 0 \qquad \blacksquare$$

§ **Problem 4.4.34.** *Prove that the diagonals of the parallelogram formed by the four straight lines*
$$\sqrt{3}x + y = 0, \quad \sqrt{3}y + x = 0, \quad \sqrt{3}x + y = 1 \text{ and } \sqrt{3}y + x = 1$$
are at right angles to one another. ◊

§§ **Solution.** The equations of the four sides are
$$\sqrt{3}x + y = 0, \qquad (4.93)$$
$$\sqrt{3}y + x = 0, \qquad (4.94)$$
$$\sqrt{3}x + y = 1, \qquad (4.95)$$
$$\sqrt{3}y + x = 1. \qquad (4.96)$$

$(4.93) - (4.94) \implies$
$$\sqrt{3}(x - y) + (y - x) = 0$$
$$\therefore (y - x)(\sqrt{3} - 1) = 0$$
$$\therefore y - x = 0$$
$$\therefore y = x.$$

$(4.95) - (4.96) \implies$
$$\sqrt{3}(x - y) + (y - x) = 0$$
$$\therefore (y - x)(\sqrt{3} - 1) = 0$$
$$\therefore y - x = 0$$
$$\therefore y = x.$$

It is clear that $y = x$ is the equation of one of the diagonals and its slope is 1.

$(4.93) + (4.96) \implies$
$$\sqrt{3}(x + y) + (y + x) = 1$$
$$\therefore (y + x)(\sqrt{3} + 1) = 1$$
$$\therefore y = -x + \frac{1}{\sqrt{3} + 1}.$$

$(4.94) + (4.95) \implies$
$$\sqrt{3}(x + y) + (y + x) = 1$$
$$\therefore (y + x)(\sqrt{3} + 1) = 1$$
$$\therefore y = -x + \frac{1}{\sqrt{3} + 1}.$$

It is clear that $y = -x + \frac{1}{\sqrt{3} + 1}$. is the equation of the other diagonal and its slope is -1, which is perpendicular to the first diagonal $y = x$.

Hence the diagonals are at right angles to one other.

4.4. Bisectors of angles between straight lines

Alternatively, we can find the coordinates of the corners of the parallelogram by solving these four equations of the sides, then find the slopes of the diagonals. ∎

§ Problem 4.4.35. *Prove the same property for the parallelogram whose sides are*
$$\frac{x}{a}+\frac{y}{b}=1,\ \frac{x}{b}+\frac{y}{a}=1,\ \frac{x}{a}+\frac{y}{b}=2,\ and\ \frac{x}{b}+\frac{y}{a}=2.$$
◊

§§ Solution. As per the approach adopted in solving the previous question (34), subtracting the first two equations and the last two equations yields the equation of the first diagonal as $y = x$.

Similarly adding first equation to the fourth one and the second equation to the third one yields the equation of the second diagonal as $y = -x + \dfrac{3ab}{a+b}$.

Hence the diagonals are at right angles to one other.

Alternatively, we can find the coordinates of the corners of the parallelogram by solving these four equations of the sides, then find the slopes of the diagonals. ∎

§ Problem 4.4.36. *One side of a square is inclined to the axis of x at an angle α and one of its extremities is at tho origin ; prove that the equations to its diagonals are*
$$y(\cos\alpha - \sin\alpha) = x(\sin\alpha + \cos\alpha)\ and$$
$$y(\sin\alpha + \cos\alpha) + x(\cos\alpha - \sin\alpha) = a.$$
Find the equations to the straight lines bisecting the angles between the following pairs of straight lines, placing first the bisector of the angle in which the origin lies. ◊

§§ Solution. Let $ABCD$ be the square such that the angular point $A(0,0)$ be the origin and the side AB be inclined to the axis of x at an angle α.

Hence the diagonal AC is inclined at $45° + \alpha$ to the x-axis. ∴ its slope $= \tan(45° + \alpha)$.

Hence the equation of the diagonal AC passing through the point $A(0,0)$ is
$$y - 0 = \tan(45° + \alpha)(x - 0)$$
$$\therefore y = \frac{\tan 45° + \tan\alpha}{1 - \tan 45° \cdot \tan\alpha}x$$
$$\therefore y = \frac{1 + \tan\alpha}{1 - \tan\alpha}x = \frac{\cos\alpha + \sin\alpha}{\cos\alpha - \sin\alpha}x$$
$$\therefore y(\cos\alpha - \sin\alpha) = x(\sin\alpha + \cos\alpha).$$

It is easy to see that the length of the perpendicular from the origin $A(0,0)$ to the other diagonal BD is $\dfrac{a}{\sqrt{2}}$ and the perpendicular is inclined at $45° + \alpha$ to the x-axis.

Hence, by *Art.* 53, the equation of the diagonal BD is
$$x\cos(45° + \alpha) + y\sin(45° + \alpha) = \frac{a}{\sqrt{2}}$$
$$\therefore x(\frac{1}{\sqrt{2}}\cos\alpha - \frac{1}{\sqrt{2}}\sin\alpha) + y(\frac{1}{\sqrt{2}}\cos\alpha + \frac{1}{\sqrt{2}}\sin\alpha) = \frac{a}{\sqrt{2}}$$
$$\therefore y(\sin\alpha + \cos\alpha) + x(\cos\alpha - \sin\alpha) = a.$$

4.4. Bisectors of angles between straight lines

Alternative Solution : The coordinates of the angular point B are $(a\cos\alpha, a\sin\alpha)$.

Slope of the diagonal BD, which is perpendicular to the diagonal AC is $= -\dfrac{1}{\tan(45° + \alpha)} = -\dfrac{\cos\alpha - \sin\alpha}{\cos\alpha + \sin\alpha}$.

Hence the equation of the diagonal BD is
$$y - a\sin\alpha = -\dfrac{\cos\alpha - \sin\alpha}{\cos\alpha + \sin\alpha}(x - a\cos\alpha)$$
$$y(\cos\alpha + \sin\alpha) + x(\cos\alpha - \sin\alpha) = a(\sin^2\alpha + \cos^2\alpha) = a$$
$$\therefore y(\sin\alpha + \cos\alpha) + x(\cos\alpha - \sin\alpha) = a. \qquad \blacksquare$$

Find the equations to the straight lines bisecting the angles between the following pairs of straight lines, placing first the bisector of the angle in which the origin lies.

§ Problem 4.4.37. $x + y\sqrt{3} = 6 + 2\sqrt{3}$ and $x - y\sqrt{3} = 6 - 2\sqrt{3}$. ◊

§§ Solution. Let us rewrite the equations to the straight lines in such a way that the constant terms are negative. Following this, the equations to the bisectors are:
$$\dfrac{x + y\sqrt{3} - 6 - 2\sqrt{3}}{\sqrt{1^2 + (\sqrt{3})^2}} = \dfrac{\pm(x - y\sqrt{3} - 6 + 2\sqrt{3})}{\sqrt{1^2 + (-\sqrt{3})^2}}$$
$$\therefore x + y\sqrt{3} - 6 - 2\sqrt{3} = \pm(x - y\sqrt{3} - 6 + 2\sqrt{3})$$

Taking the positive sign, we get the equation of the bisector of the angle in which the origin lies as follows:
$$2\sqrt{3}y = 4\sqrt{3}$$
$$\therefore y = 2.$$

Taking the negative sign, we get the equation of the other bisector as follows:
$$2x = 12$$
$$\therefore x = 6. \qquad \blacksquare$$

§ Problem 4.4.38. $12x + 5y - 4 = 0$ and $3x + 4y + 7 = 0$. ◊

§§ Solution. Let us rewrite the equations to the straight lines in such a way that the constant terms are negative. Following this, the equations to the bisectors are:
$$\dfrac{12x + 5y - 4}{\sqrt{12^2 + 5^2}} = \pm\left(\dfrac{-3x - 4y - 7}{\sqrt{(-3)^2 + (-4)^2}}\right)$$
$$\therefore \dfrac{12x + 5y - 4}{13} = \pm\left(\dfrac{-3x - 4y - 7}{5}\right)$$
$$\therefore 60x + 25y - 20 = \pm(-39x - 52y - 91)$$

Hence the equations are:
$$99x + 77y + 71 = 0$$
and
$$21x - 27y - 111 = 0$$
$$\therefore 7x - 9y - 37 = 0. \qquad \blacksquare$$

§ Problem 4.4.39. $4x + 3y - 7 = 0$ and $24x + 7y - 31 = 0$. ◊

4.4. Bisectors of angles between straight lines

§§ Solution. Let us rewrite the equations to the straight lines in such a way that the constant terms are negative. Following this, the equations to the bisectors are:

$$\frac{4x+3y-7}{\sqrt{4^2+3^2}} = \pm\left(\frac{24x+7y-31}{\sqrt{24^2+7^2}}\right)$$

$$\therefore \frac{4x+3y-7}{5} = \pm\left(\frac{24x+7y-31}{25}\right)$$

$$\therefore 20x+15y-35 = \pm(24x+7y-31)$$

Hence the equations are:

$$4x-8y+4 = 0$$
$$\therefore x-2y+1 = 0.$$

and

$$44x+22y = 66$$
$$\therefore 2x+y-3 = 0.$$ ∎

§ Problem 4.4.40. $2x+y=4$ and $y+3x=5$. ◊

§§ Solution. Let us rewrite the equations to the straight lines in such a way that the constant terms are negative. Following this, the equations to the bisectors are:

$$\frac{2x+y-4}{\sqrt{2^2+1^2}} = \pm\left(\frac{y+3x-5}{\sqrt{1^2+3^2}}\right)$$

$$\therefore \frac{2x+y-4}{\sqrt{5}} = \pm\left(\frac{y+3x-5}{\sqrt{10}}\right)$$

$$\therefore \sqrt{2}(2x+y-4) = \pm(y+3x-5)$$

Hence the equations are

$$\therefore x(2\sqrt{2}-3)+y(\sqrt{2}-1) = 4\sqrt{2}-5$$

and

$$\therefore x(2\sqrt{2}+3)+y(\sqrt{2}+1) = 4\sqrt{2}+5$$ ∎

§ Problem 4.4.41. $y-b = \dfrac{2m}{1-m^2}(x-a)$ and $y-b = \dfrac{2m'}{1-m'^2}(x-a)$. ◊

§§ Solution. It is easy to decipher from the given equation that the first line passes through the point (a,b) and its slope $= \dfrac{2m}{1-m^2}$.

Let us denote m by $\tan\theta_1$, then slope of the first line becomes

$$\frac{2\tan\theta_1}{1-\tan^2\theta_1} = \tan 2\theta_1$$

Hence the first line is inclined at $2\theta_1$ to the x-axis.

Similarly, it is easy to decipher from the given equation that the second line passes through the point (a,b) and denoting m' by $\tan\theta_2$, its slope becomes

$$\frac{2m'}{1-m'^2} = \frac{2\tan\theta_2}{1-\tan^2\theta_2} = \tan 2\theta_2$$

Hence the second line is inclined at $2\theta_2$ to the x-axis.

Since both the lines pass through the point $A(a,b)$, let the two straight lines be $AL1$ and $AL2$, and let the bisectors of the angles between them be $AM1$ and $AM2$. Obviously, these bisectors also pass through the point $A(a,b)$.

It is easy to see from the diagram that the following is true for the bisector $AM1$ of the interior angle $\angle L1AL2$

For $\triangle AM1L2 : 2\theta_1 = 2\theta_2 + 2\alpha$, and for $\triangle AL1L2 : \theta = 2\theta_2 + \alpha$

4.4. Bisectors of angles between straight lines

$$\therefore 2\theta = 2\theta_1 + 2\theta_2$$
$$\therefore \theta = \theta_1 + \theta_2$$

Hence slope of the bisector $AM1$ is
$$\tan\theta = \tan(\theta_1 + \theta_2) = \frac{\tan\theta_1 + \tan\theta_2}{1 - \tan\theta_1 \cdot \tan\theta_2} = \frac{m + m'}{1 - mm'}$$

The required equation of the bisector $AM1$ passing through the point $A(a, b)$ with slope as $\dfrac{m + m'}{1 - mm'}$ is given by
$$y - b = \frac{m + m'}{1 - mm'}(x - a)$$
$$\therefore (y - b)(1 - mm') - (x - a)(m + m') = 0.$$

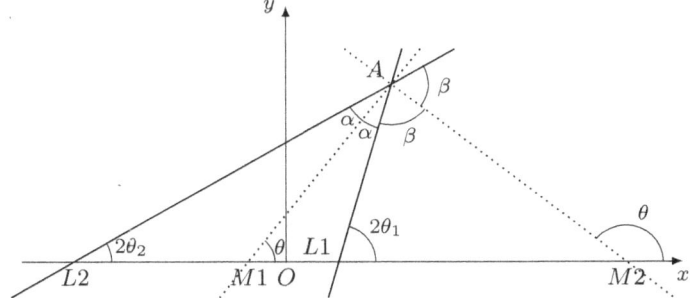

Similarly the following is true for the bisector $AM2$ of the exterior angle

For $\triangle AL1L2$: $2\theta_1 = 2\theta_2 + \pi - 2\beta$, and for $\triangle AL1M2$: $\theta = 2\theta_1 + \beta$
$$\therefore 2\left(\frac{\pi}{2} - \theta\right) = 2\theta_1 + 2\theta_2$$
$$\therefore \theta = \frac{\pi}{2} + \theta_1 + \theta_2.$$

Hence slope of the bisector $AM2$ is
$$\tan\theta = \tan\left(\frac{\pi}{2} + \theta_1 + \theta_2\right) = -\cot(\theta_1 + \theta_2) = -\frac{1}{\tan(\theta_1 + \theta_2)}$$
$$= -\frac{1}{\frac{\tan\theta_1 + \tan\theta_2}{1 - \tan\theta_1 \cdot \tan\theta_2}} = -\frac{1 - mm'}{m + m'}.$$

The required equation of the bisector $AM2$ passing through the point $A(a, b)$ with slope as $-\dfrac{1 - mm'}{m + m'}$ is given by
$$y - b = -\frac{1 - mm'}{m + m'}(x - a)$$
$$\therefore (y - b)(m + m') + (x - a)(1 - mm') = 0.$$

Alternative Solution :
Then given equations are
$$y - \frac{2m}{1 - m^2}x - b + \frac{2ma}{1 - m^2} = 0, \text{ and}$$
$$y - \frac{2m'}{1 - m'^2}x - b + \frac{2m'a}{1 - m'^2} = 0.$$

4.4. Bisectors of angles between straight lines

Hence the equations of the bisectors are

$$\frac{-\frac{2m}{1-m^2}x + y - b + \frac{2ma}{1-m^2}}{\sqrt{\frac{4m^2}{(1-m^2)^2}+1}} = \pm \frac{-\frac{2m'}{1-m'^2}x + y - b + \frac{2m'a}{1-m'^2}}{\sqrt{\frac{4m'^2}{(1-m'^2)^2}+1}}$$

$$\therefore \frac{(y-b)(1-m^2) - 2mx + 2ma}{1+m^2} = \pm \frac{(y-b)(1-m'^2) - 2m'x + 2m'a}{1+m'^2}$$

$$\therefore \left[(y-b)(1-m^2) - 2m(x-a)\right](1+m'^2)$$
$$= \pm \left[(y-b)(1-m'^2) - 2m'(x-a)\right](1+m^2) \qquad (4.97)$$

Taking the positive sign, the equation of the bisector becomes

$$\left[(y-b)(1-m^2) - 2m(x-a)\right](1+m'^2)$$
$$= \left[(y-b)(1-m'^2) - 2m'(x-a)\right](1+m^2)$$

$$\therefore (y-b)\left[(1+m'^2 - m^2 - m^2m'^2) - (1 - m'^2 + m^2 - m^2m'^2)\right]$$
$$-(x-a)\left[(2m + 2mm'^2) - (2m' + 2m'm^2)\right] = 0$$

$$\therefore (y-b)(2m'^2 - 2m^2) - (x-a)\left[2(m-m') + 2mm'(m'-m)\right] = 0$$

$$\therefore 2(y-b)(m'-m)(m'+m) - 2(x-a)(m'-m)(mm'-1) = 0$$

$$\therefore (y-b)(m+m') + (x-a)(1-mm') = 0.$$

Taking the negative sign, the equation of the bisector becomes

$$\left[(y-b)(1-m^2) - 2m(x-a)\right](1+m'^2)$$
$$= -\left[(y-b)(1-m'^2) - 2m'(x-a)\right](1+m^2)$$

$$\therefore (y-b)\left[(1+m'^2 - m^2 - m^2m'^2) + (1 - m'^2 + m^2 - m^2m'^2)\right]$$
$$-(x-a)\left[(2m + 2mm'^2) + (2m' + 2m'm^2)\right] = 0$$

$$\therefore (y-b)(2 - 2m^2m'^2) - (x-a)\left[2(m+m') + 2mm'(m'+m)\right] = 0$$

$$\therefore 2(y-b)(1+mm')(1-mm') - 2(x-a)(m'+m)(mm'+1) = 0$$

$$\therefore (y-b)(1-mm') - (x-a)(m+m') = 0. \qquad \blacksquare$$

Find the equations to the bisectors of the internal angles of the triangles the equations of whose sides are respectively

§ **Problem 4.4.42.** $3x + 4y = 6$, $12x - 5y = 3$ and $4x - 3y + 12 = 0$. ◊

§§ **Solution.** Let the given equations denote the sides AB, BC, and CA respectively:

$$3x + 4y = 6 \qquad (4.98)$$
$$12x - 5y = 3 \qquad (4.99)$$
$$4x - 3y + 12 = 0 \qquad (4.100)$$

Solving the equations (4.98) : AB and (4.99) : BC, we get the coordinates of the angular point B :

(4.98) × 4 − (4.99) $\implies$

$$21y = 21, \therefore y = 1$$
$$\implies 3x = 6 - 4, \therefore x = \frac{2}{3}$$

$\therefore$ the coordinates of the angular point B is $\left(\frac{2}{3}, 1\right)$.

Solving the equations (4.98) : AB and (4.100) : CA, we get the coordinates of the angular point A :

4.4. Bisectors of angles between straight lines

$(4.98) \times 3 + (4.99) \times 4 \implies$
$$25x = 18 - 48 = -30, \therefore x = -\frac{6}{5}$$
$$\implies y = \frac{6-3x}{4} = \frac{48}{20} = \frac{12}{5}$$

∴ the coordinates of the angular point A is $\left(-\frac{6}{5}, \frac{12}{5}\right)$.

Solving the equations (4.99) : BC and (4.100) : CA, we get the coordinates of the angular point C :

$(4.99) - (4.100) \times 3 \implies$
$$4y = 39, \therefore y = \frac{39}{4}$$
$$\implies 4x = 3y - 12 = \frac{69}{4}, \therefore x = \frac{69}{16}$$

∴ the coordinates of the angular point C is $\left(\frac{69}{16}, \frac{39}{4}\right)$.

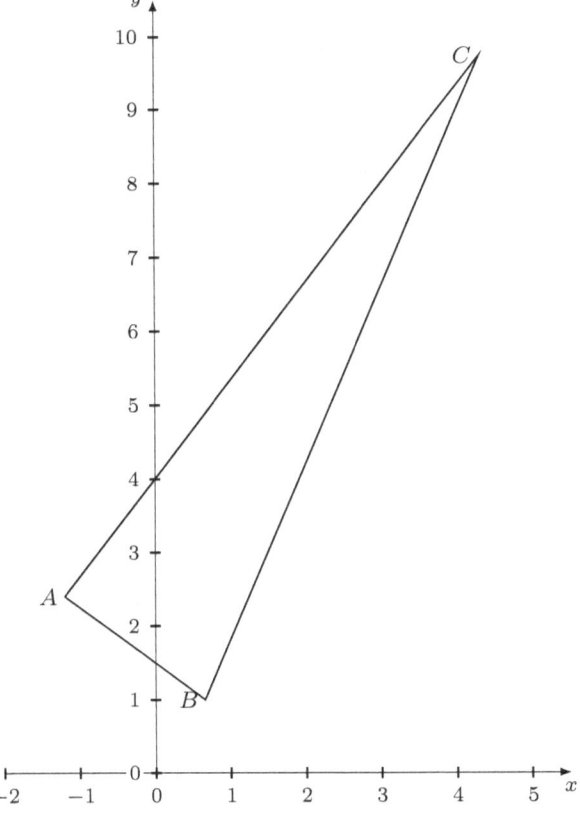

Let us rewrite the equations of AB, BC, and CA such that the constant terms are negative :

4.4. Bisectors of angles between straight lines

$$AB : 3x + 4y - 6 = 0$$
$$BC : 12x - 5y - 3 = 0$$
$$CA : 3y - 4x - 12 = 0$$

It is clear from the figure that the origin lies outside the interior angle between AB and BC, hence from Art. 84, the bisector of the interior angle $\angle ABC$ is

$$\frac{3x + 4y - 6}{\sqrt{3^2 + 4^2}} = -\frac{12x - 5y - 3}{\sqrt{12^2 + (-5)^2}}$$

$$\therefore \frac{3x + 4y - 6}{5} = -\frac{12x - 5y - 3}{13}$$

$$\therefore 39x + 52y - 78 = -60x + 25y + 15$$

$$\therefore 99x + 27y = 93$$

$$\therefore 33x + 9y = 31.$$

It is clear from the figure that the origin lies inside the interior angle between BC and CA, hence from Art. 84, the bisector of the interior angle $\angle BCA$ is

$$\frac{12x - 5y - 3}{\sqrt{12^2 + (-5)^2}} = +\frac{3y - 4x - 12}{\sqrt{3^2 + (-4)^2}}$$

$$\therefore \frac{12x - 5y - 3}{13} = +\frac{3y - 4x - 12}{5}$$

$$\therefore 60x - 25y - 15 = 39y - 52x - 156$$

$$\therefore 112x - 64y + 141 = 0.$$

It is clear from the figure that the origin lies outside the interior angle between AB and AC, hence from Art. 84, the bisector of the interior angle $\angle BAC$ is

$$\frac{3x + 4y - 6}{\sqrt{3^2 + 4^2}} = -\frac{3y - 4x - 12}{\sqrt{3^2 + (-4)^2}}$$

$$\therefore 3x + 4y - 6 = -(3y - 4x - 12)$$

$$\therefore 7y - x = 18. \qquad \blacksquare$$

§ Problem 4.4.43. $3x + 5y = 15$, $x + y = 4$ and $2x + y = 6$. ◊

§§ Solution. Let the given equations denote the sides AB, BC, and CA respectively:

$$3x + 5y = 15 \qquad (4.101)$$
$$x + y = 4 \qquad (4.102)$$
$$2x + y = 6 \qquad (4.103)$$

Solving the equations (4.101) : AB and (4.102) : BC, we get the coordinates of the angular point B :

$(4.101) - (4.102) \times 3 \implies$

$$2y = 3, \therefore y = \frac{3}{2}$$

$$\implies x = 4 - y = 4 - \frac{3}{2}, \therefore x = \frac{5}{2}$$

$\therefore$ the coordinates of the angular point B is $\left(\frac{5}{2}, \frac{3}{2}\right)$.

Solving the equations (4.101) : AB and (4.103) : CA, we get the coordinates of the angular point A :

4.4. Bisectors of angles between straight lines

$(4.101) - (4.102) \times 5 \implies$
$$-7x = -15, \therefore x = \frac{15}{7}$$
$$\implies y = 6 - 2x = \frac{12}{7}$$

∴ the coordinates of the angular point A is $\left(\dfrac{15}{7}, \dfrac{12}{7}\right)$.

Solving the equations (4.102) : BC and (4.103) : CA, we get the coordinates of the angular point C :

$(4.102) - (4.103) \implies$
$$-x = -2, \therefore x = 2$$
$$\implies y = 2$$

∴ the coordinates of the angular point C is $(2, 2)$.

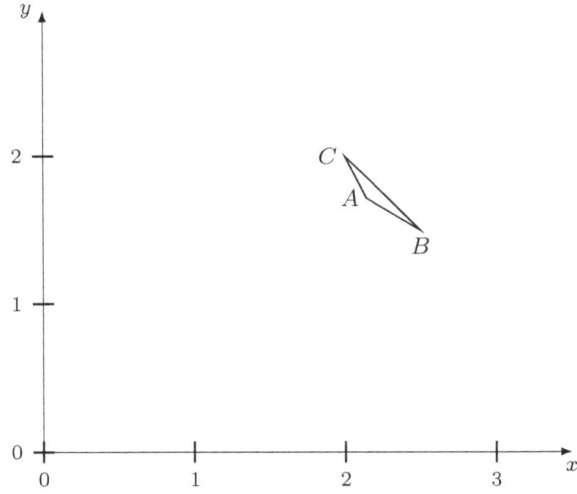

Hence the internal bisectors are as follows:

$$\frac{3x + 5y - 15}{\sqrt{34}} = +\frac{2x + y - 6}{\sqrt{5}},$$
$$\frac{3x + 5y - 15}{\sqrt{34}} = -\frac{x + y - 4}{\sqrt{2}},$$
$$\frac{2x + y - 6}{\sqrt{5}} = -\frac{x + y - 4}{\sqrt{2}}$$

Upon simplification, the above becomes as follows:

$$x(3 + \sqrt{17}) + y(5 + \sqrt{17}) = 15 + 4\sqrt{17};$$
$$x(4 + \sqrt{10}) + y(2 + \sqrt{10}) = 4\sqrt{10} + 12;$$
$$x(2\sqrt{34} - 3\sqrt{5}) + y(\sqrt{34} - 5\sqrt{5}) = 6\sqrt{34} - 15\sqrt{5}. \quad \blacksquare$$

§ Problem 4.4.44. *Find the equations to the straight lines passing through the foot of the perpendicular from the point (h, k) upon the*

4.4. Bisectors of angles between straight lines

straight line $Ax + By + C = 0$ and bisecting the angles between the perpendicular and the given straight line. ◊

§§ Solution. Slope of the line $Ax + By + C = 0$ is $-\dfrac{A}{B}$.

Hence the slope of the line perpendicular to the above line is $\dfrac{B}{A}$.

Hence the equation of the line perpendicular to the line $Ax+By+C=0$ and passing through the point (h, k) is:
$$y - k = \frac{B}{A}(x - h)$$
$$\therefore A(y - k) - B(x - h) = 0.$$

Alternatively, by *Art.* 70., the equation of the line perpendicular to $Ax + By + C = 0$ is
$$Ay - Bx + C_1 = 0$$
Since it passes through (h, k), $\therefore C_1 = Bh - Ak$.
Hence the line becomes
$$Ay - Bx + Bh - Ak = 0$$
$$\therefore A(y - k) - B(x - h) = 0.$$

Hence by *Art.* 84., the equation of the bisectors are
$$\frac{A(y - k) - B(x - h)}{\sqrt{A^2 + B^2}} = \pm \frac{Ax + By + C}{\sqrt{A^2 + B^2}}$$
$$\therefore A(y - k) - B(x - h) = \pm(Ax + By + C). \blacksquare$$

§ Problem 4.4.45. *Find the direction in which a straight line must be drawn through the point $(1, 2)$, so that its point of intersection with the line $x + y = 4$ may be at a distance $\dfrac{\sqrt{6}}{3}$ from this point.* ◊

§§ Solution. Let the line be inclined at an angle θ to the axis of x, then by Art. 86, the following point must lie on the line $x + y = 4$,
$$\left(1 + \frac{\sqrt{6}}{3} \cos\theta,\ 2 + \frac{\sqrt{6}}{3} \sin\theta \right)$$
i.e.,
$$1 + \frac{\sqrt{6}}{3} \cos\theta + 2 + \frac{\sqrt{6}}{3} \sin\theta = 4$$
$$\therefore \cos\theta + \sin\theta = \frac{3}{\sqrt{6}} = \frac{\sqrt{3}}{\sqrt{2}}$$
$$\therefore \frac{1}{\sqrt{2}} \cos\theta + \frac{1}{\sqrt{2}} \sin\theta = \frac{\sqrt{3}}{2}$$
$$\therefore \sin(45° + \theta) = \frac{\sqrt{3}}{2}$$
$$\therefore 45° + \theta = 60°, \text{ or } 120°$$
$$\therefore \theta = 15°, \text{ or } 75°. \blacksquare$$

Chapter 5

The Straight Line : Polar Equations, Oblique Coordinates and Loci

5.1 Oblique Coordinates

§ Problem 5.1.1. *The axes being inclined at an angle of* $60°$*, find the inclination to the axis of* x *of the straight lines whose equations are*

(1) $y = 2x + 5$

(2) $2y = \left(\sqrt{3} - 1\right)x + 7.$ ◇

§§ Solution. (1) By *Art.* 91, the angle is
$$= \tan^{-1} \frac{m \sin \omega}{1 + m \cos \omega}$$
$$= \frac{2 \cdot \sin 60°}{1 + 2 \cos 60°} = \tan^{-1} \frac{\sqrt{3}}{2}.$$

(2) By *Art.* 91, the angle is

$$= \tan^{-1} \frac{m \sin \omega}{1 + m \cos \omega} = \tan^{-1} \frac{\dfrac{\sqrt{3} - 1}{2} \cdot \sin 60°}{1 + \dfrac{\sqrt{3} - 1}{2} \cos 60°}$$

$$= \tan^{-1} \frac{\dfrac{\sqrt{3} - 1}{2} \cdot \dfrac{\sqrt{3}}{2}}{1 + \dfrac{\sqrt{3} - 1}{2} \cdot \dfrac{1}{2}} = \tan^{-1} \frac{\sqrt{3}(\sqrt{3} - 1)}{3 + \sqrt{3}} = \tan^{-1} \frac{\sqrt{3} - 1}{\sqrt{3} + 1}$$

5.1. Oblique Coordinates

$$= \tan^{-1} \frac{1 - \frac{1}{\sqrt{3}}}{1 + 1 \cdot \frac{1}{\sqrt{3}}} = \tan^{-1} \tan(45° - 30°) = 15°.$$ ∎

§ Problem 5.1.2. *The axes being inclined at an angle of* $120°$, *find the tangent of the angle between the two straight lines*
$$8x + 7y = 1 \text{ and } 28x - 73y = 101.$$ ◊

§§ Solution. By *Art.* 93, the angle is

$$= \tan^{-1} \frac{(m - m') \sin \omega}{1 + (m + m') \cos \omega + mm'}$$

$$= \tan^{-1} \frac{\left(\frac{28}{73} + \frac{8}{7}\right) \sin 120°}{1 + \left(\frac{28}{73} - \frac{8}{7}\right) \cos 120° - \frac{28}{73} \cdot \frac{8}{7}}$$

$$= \tan^{-1} \frac{30\sqrt{3}}{37}.$$ ∎

§ Problem 5.1.3. *With oblique coordinates find the tangent of the angle between the straight lines*
$$y = mx + c \text{ and } my + x = d.$$ ◊

§§ Solution. By *Art.* 93, the tangent of the angle is

$$= \frac{(m - m') \sin \omega}{1 + (m + m') \cos \omega + mm'}, \text{ where } m' = -\frac{1}{m}$$

$$= \frac{\left(m + \frac{1}{m}\right) \sin \omega}{1 + \left(m - \frac{1}{m}\right) \cos \omega - m \cdot \frac{1}{m}} = \frac{(m^2 + 1)}{(m^2 - 1)} \tan \omega$$ ∎

§ Problem 5.1.4. *If* $y = x \tan \frac{11\pi}{24}$ *and* $y = x \tan \frac{19\pi}{24}$ *represent two straight lines at right angles, prove that the angle between the axes is* $\frac{\pi}{4}$. ◊

§§ Solution. By *Art.* 93, *Cor.* 2, we have
$$1 + (m + m') \cos \omega + mm' = 0$$
$$\therefore 1 + \left(\tan \frac{11\pi}{24} + \tan \frac{19\pi}{24}\right) \cos \omega + \tan \frac{11\pi}{24} \cdot \tan \frac{19\pi}{24} = 0$$
$$\therefore \cos \omega = -\frac{1 + \tan \frac{11\pi}{24} \cdot \tan \frac{19\pi}{24}}{\tan \frac{11\pi}{24} + \tan \frac{19\pi}{24}} = -\frac{\cos \frac{19\pi - 11\pi}{24}}{\sin \frac{19\pi + 11\pi}{24}} = -\frac{\frac{1}{2}}{-\frac{1}{\sqrt{2}}} = \frac{1}{\sqrt{2}}$$
$$\therefore \omega = \frac{\pi}{4}$$ ∎

§ Problem 5.1.5. *Prove that the straight lines* $y + x = c$ *and* $y = x + d$ *are at right angles, whatever be the angle between the axes.* ◊

§§ Solution. Let us evaluate the following expression
$$1 + (m + m') \cos \omega + mm', \text{ where } m = -1, \text{ and } m = 1$$
$$= 1 + (-1 + 1) \cos \omega - 1 = 0, \text{ irrespective of the value of } \omega.$$

Hence by *Art.* 93, *Cor.* 2, the given lines are at right angles, whatever be the angle between the axes. ∎

5.1. Oblique Coordinates

§ Problem 5.1.6. *Prove that the equation to the straight line which passes through the point (h, k) and is perpendicular to the axis of x is*
$$x + y \cos \omega = h + k \cos \omega. \qquad \diamond$$

§§ Solution. The equation of the line with slope m and passing through the point (h, k) is
$$y - k = m(x - h)$$
The equation of the axis of x is $y = 0$. Its slope is $m' = 0$.
Since the above two lines are perpendicular to each other, hence by Art. 93, Cor. 2,
$$1 + (m + m') \cos \omega + mm' = 0, \text{ where } m' = 0$$
$$\therefore 1 + m \cos \omega = 0$$
$$\therefore m = -\frac{1}{\cos \omega}.$$
Hence the required equation becomes
$$y - k = -\frac{1}{\cos \omega}(x - h)$$
$$\therefore x + y \cos \omega = h + k \cos \omega. \qquad \blacksquare$$

§ Problem 5.1.7. *Find the equations to the sides and diagonals of a regular hexagon, two of its sides, which meet in a corner, being the axes of coordinates.* $\diamond$

§§ Solution. Let $OABCDE$ denote the regular hexagon, two of its sides : OA and OE, which meet in a corner : origin O, being the axes of coordinates, i.e., OA is the x-axis and OE is the y-axis.

Hence the equation of OA, which is the x-axis, is $y = 0$.
Similarly, the equation of OE, which is the y-axis, is $x = 0$.
Length of the x-intercept and y-intercept by the line AB are a and $-a$ respectively.
Hence its equation is given by, Art. 50,
$$\frac{x}{a} + \frac{y}{b} = 1$$
$$\therefore \frac{x}{a} + \frac{y}{-a} = 1$$
$$\therefore x - y = a$$
Hence the equation of AB is $y = x - a$.
Length of the x-intercept and y-intercept by the line ED are $-a$ and a respectively.
Hence its equation is given by, Art. 50,
$$\frac{x}{a} + \frac{y}{b} = 1$$
$$\therefore \frac{x}{-a} + \frac{y}{a} = 1$$
$$\therefore y - x = a$$
Hence the equation of ED is $y = x + a$.
The sides CD and BE are parallel to the axis of x and their ordinates are $2a$ and a respectively.
Hence, the equation of CD is $y = 2a$.
Similarly, the equation of BE is $y = a$.
The sides CB and AD are parallel to the axis of y and their abscissae are $2a$ and a respectively.
Hence the equation of CB is $x = 2a$.

5.1. Oblique Coordinates

Similarly, the equation of AD is $x = a$.

The diagonal OC bisects the angle between the axes, hence its equation is $y = x$. ∎

§ Problem 5.1.8. *From each corner of a parallelogram a perpendicular is drawn upon the diagonal which does not pass through that corner and these are produced to form another parallelogram ; show that its diagonals are perpendicular to the sides of the first parallelogram and that they both have the same center.* ◊

§§ Solution. Let $ABCD$ be the parallelogram and let its diagonals CA and DB be the x-axis and y-axis respectively such that the angular points are $A(a, 0)$, $C(-a, 0)$, $B(0, b)$ and $D(0, -b)$. respectively.

Equation of the line with slope m and passing through the angular point $B(0, b)$ is

$$y - b = m(x - 0)$$
$$\therefore y = mx + b$$

If the above line is perpendicular to the diagonal CA, which is the x-axis, i.e., $y = 0$, then by Art. 93, Cor. 2,

$$1 + (m + m')\cos\omega + mm' = 0, \text{ where } m' = 0$$
$$\therefore 1 + m\cos\omega = 0$$
$$\therefore m = -\frac{1}{\cos\omega}.$$

Hence the equation of the side of the second parallelogram passing through the angular point $B(0, b)$ and perpendicular to the diagonal CA becomes

$$y = -\frac{1}{\cos\omega}x + b$$

$$\therefore x + y\cos\omega = b\cos\omega. \tag{5.1}$$

Similarly the equation of the side of the second parallelogram passing through the angular point $D(0, -b)$ and perpendicular to the diagonal CA is

$$\therefore x + y\cos\omega = -b\cos\omega. \tag{5.2}$$

Alternatively, by *Art.* 91., the equation of a line inclined to x-axis at an angle θ with y-intercept as c is given by:

$$y = x\frac{\sin\theta}{\sin(\omega - \theta)} + c \tag{5.3}$$

Hence, equation of the lines, perpendicular to the x-axis ($\theta = 90°$), i.e., the diagonal CA is given by

$$y = x\frac{\sin 90°}{\sin(\omega - 90°)} + c = -\frac{x}{\cos\omega} + c$$

i.e.,

$$x + y\cos\omega = c \tag{5.4}$$

Similarly, equation of the lines, perpendicular to the y-axis ($\theta = 90° + \omega$), i.e., the diagonal DB is given by

$$y = x\frac{\sin(90° + \omega)}{\sin(\omega - 90° - \omega)} + c = -x\cos\omega + c$$

i.e.,

$$x\cos\omega + y = c \tag{5.5}$$

Hence, equation of the side of the second parallelogram passing through the point $B(0, b)$ and perpendicular to the diagonal CA of

5.1. Oblique Coordinates

the first parallelogram can be obtained by substituting $B(0,b)$ in the equation (5.4) as follows:
$$0 + b\cos\omega = c, \implies c = b\cos\omega$$
Hence the required equation is
$$x + y\cos\omega = b\cos\omega \tag{5.6}$$
Similarly, equation of the side of the second parallelogram passing through the point $D(0,-b)$ and perpendicular to the diagonal CA of the first parallelogram can be obtained by substituting $D(0,-b)$ in the equation (5.4) as follows:
$$0 + -b\cos\omega = c, \implies c = -b\cos\omega$$
Hence the required equation is
$$x + y\cos\omega = -b\cos\omega \tag{5.7}$$
Similarly, equation of the side of the second parallelogram passing through the point $A(a,0)$ and perpendicular to the diagonal DB of the first parallelogram can be obtained by substituting $A(a,0)$ in the equation (5.5) as follows:
$$a\cos\omega + 0 = c, \implies c = a\cos\omega$$
Hence the required equation is
$$x\cos\omega + y = a\cos\omega \tag{5.8}$$
Similarly, equation of the side of the second parallelogram passing through the point $C(-a,0)$ and perpendicular to the diagonal DB of the first parallelogram can be obtained by substituting $C(-a,0)$ in the equation (5.5) as follows:
$$-a\cos\omega + 0 = c, \implies c = -a\cos\omega$$
Hence the required equation is
$$x\cos\omega + y = -a\cos\omega \tag{5.9}$$
The equation of a line passing through the intersection of (5.6) and (5.8) is given by
(5.6) $\times\, a\, -\,$ (5.8) $\times\, b \implies$
$$x(a - b\cos\omega) + y(a\cos\omega - b) = 0 \tag{5.10}$$
The equation of a line passing through the intersection of (5.7) and (5.9) is given by
(5.7) $\times\, a\, -\,$ (5.9) $\times\, b \implies$
$$x(a - b\cos\omega) + y(a\cos\omega - b) = 0 \tag{5.11}$$
But the equations (5.10) and (5.11) are the same, hence (5.10) is the equation of one of the diagonals of the second parallelogram.

Similarly, the equation of the other diagonal of the second parallelogram is given by the finding that the equation of the line passing through the intersection of (5.6) and (5.9) as well as the intersection of (5.7) and (5.8) is the same line :
(5.6) $\times\, a\, +\,$ (5.9) $\times\, b$ and (5.7) $\times\, a\, +\,$ (5.8) $\times\, b \implies$
$$x(a + b\cos\omega) + y(a\cos\omega + b) = 0 \tag{5.12}$$
It is easy to see that both the diagonals (5.10) and (5.11) of the second parallelogram pass through the origin$(0,0)$, hence both the parallelograms have the same centre, which is the origin.

Slope of the diagonal (5.10) is
$$m = -\frac{(a - b\cos\omega)}{(a\cos\omega - b)}$$

5.1. Oblique Coordinates

Equation of the side AB of the first parallelogram is $\dfrac{x}{a} + \dfrac{y}{b} = 1$, and its slope is
$$m' = -\frac{b}{a}$$

By *Art*. 93, *Cor*. 2, let us evaluate the expression
$1 + (m + m')\cos\omega + mm'$, where $m = -\dfrac{(a - b\cos\omega)}{(a\cos\omega - b)}$ and $m' = -\dfrac{b}{a}$

$= 1 + \left(-\dfrac{(a - b\cos\omega)}{(a\cos\omega - b)} - \dfrac{b}{a}\right)\cos\omega + \left(-\dfrac{(a - b\cos\omega)}{(a\cos\omega - b)} \cdot -\dfrac{b}{a}\right)$

$= 1 - \left(\dfrac{(a^2 - ab\cos\omega + ab\cos\omega - b^2)\cos\omega - ab + b^2\cos\omega}{a^2\cos\omega - ab}\right)$

$= 1 - \left(\dfrac{a^2\cos\omega - ab}{a^2\cos\omega - ab}\right) = 0$

Hence *Art*. 93, *Cor*. 2, the diagonal (5.10) of the second parallelogram is perpendicular to the side AB of the first parallelogram.

Slope of the diagonal (5.12) is
$$m = -\frac{(a + b\cos\omega)}{(a\cos\omega + b)}$$

Equation of the side AD of the first parallelogram is $\dfrac{x}{a} - \dfrac{y}{b} = 1$, and its slope is
$$m' = \frac{b}{a}$$

By *Art*. 93, *Cor*. 2, let us evaluate the expression
$1 + (m + m')\cos\omega + mm'$, where $m = -\dfrac{(a + b\cos\omega)}{(a\cos\omega + b)}$ and $m' = \dfrac{b}{a}$

$= 1 + \left(-\dfrac{(a + b\cos\omega)}{(a\cos\omega + b)} + \dfrac{b}{a}\right)\cos\omega + \left(-\dfrac{(a + b\cos\omega)}{(a\cos\omega + b)} \cdot \dfrac{b}{a}\right)$

$= 1 + \left(\dfrac{(-a^2 - ab\cos\omega + ab\cos\omega + b^2)\cos\omega - ab - b^2\cos\omega}{a^2\cos\omega + ab}\right)$

$= 1 - \left(\dfrac{a^2\cos\omega + ab}{a^2\cos\omega + ab}\right) = 0$

Hence *Art*. 93, *Cor*. 2, the diagonal (5.12) of the second parallelogram is perpendicular to the side AD of the first parallelogram. ∎

§ Problem 5.1.9. *If the straight lines $y = m_1 x + c_1$ and $y = m_2 x + c_2$ make equal angles with the axis of x and be not parallel to one another, prove that $m_1 + m_2 + 2m_1 m_2 \cos\omega = 0$.* ◊

§§ Solution. The angle made with the x-axis by the line $y = m_1 x + c_1$ is
$$\tan^{-1}\frac{m_1 \sin\omega}{1 + m_1 \cos\omega} \tag{5.13}$$

Since the second line is not parallel to the first line, hence the angle made by the second line with x-axis is supplementary, i.e.,
$$\pi - \tan^{-1}\frac{m_2 \sin\omega}{1 + m_2 \cos\omega} \tag{5.14}$$

5.1. Oblique Coordinates

Since, both the angles are equal, hence equating (5.13) to (5.14), we get the following:
$$\tan^{-1}\frac{m_1 \sin\omega}{1+m_1 \cos\omega} = \pi - \tan^{-1}\frac{m_2 \sin\omega}{1+m_2 \cos\omega}$$
$$\therefore \tan\left(\tan^{-1}\frac{m_1 \sin\omega}{1+m_1 \cos\omega}\right) = \tan\left(\pi - \tan^{-1}\frac{m_2 \sin\omega}{1+m_2 \cos\omega}\right)$$
$$\therefore \frac{m_1 \sin\omega}{1+m_1 \cos\omega} = -\frac{m_2 \sin\omega}{1+m_2 \cos\omega}$$
$$\therefore \frac{m_1}{1+m_1 \cos\omega} = -\frac{m_2}{1+m_2 \cos\omega}$$
$$\therefore m_1 + m_2 + 2m_1 m_2 \cos\omega = 0. \qquad \blacksquare$$

§ Problem 5.1.10. *The axes being inclined at an angle of* $30°$, *find the equation to the straight line which passes through the point* $(-2, 3)$ *and is perpendicular to the straight line* $y + 3x = 6$. ◊

§§ Solution. The equation of the line with slope m and passing through the point $(-2, 3)$ is
$$y - 3 = m(x+2) \qquad (5.15)$$
Since the above is perpendicular to the line $y+3x = 6$ whose slope $m' = -3$., therefore by Art. 93, Cor. 2,
$$1 + (m + m')\cos\omega + mm' = 0, \text{ where } m' = -3$$
$$\therefore 1 + (m - 3)\cos 30° - 3m = 0$$
$$\therefore m = \frac{3\sqrt{3}-2}{\sqrt{3}-6}.$$
Hence the equation (5.15) becomes
$$y(6-\sqrt{3}) + x(3\sqrt{3}-2) = 22 - 9\sqrt{3}. \qquad \blacksquare$$

§ Problem 5.1.11. *Find the length of the perpendicular drawn from the point* $(4, -3)$ *upon the straight line* $6x + 3y - 10 = 0$, *the angle between the axes being* $60°$. ◊

§§ Solution. By *Art.* 96, the length of the perpendicular is
$$= \frac{Ax' + By' + C}{\sqrt{A^2 + B^2 - 2AB\cos\omega}} \cdot \sin\omega$$
$$= \frac{6 \cdot 4 + 3 \cdot (-3) - 10}{\sqrt{6^2 + 3^2 - 2 \cdot 6 \cdot 3 \cos 60°}} \cdot \sin 60° = \frac{5}{6}. \qquad \blacksquare$$

§ Problem 5.1.12. *Find the equation to, and the length of, the perpendicular drawn from the point* $(1, 1)$ *upon the straight line* $3x + 4y + 5 = 0$, *the angle between the axes being* $120°$. ◊

§§ Solution. By *Art.* 96, the length of the perpendicular is
$$= \frac{Ax' + By' + C}{\sqrt{A^2 + B^2 - 2AB\cos\omega}} \cdot \sin\omega$$
$$= \frac{3 + 4 + 5}{\sqrt{3^2 + 4^2 - 2 \cdot 3 \cdot 4 \cos 120°}} \cdot \sin 120° = \frac{6\sqrt{111}}{37}.$$

Equation of the line with slope m and passing through the point $(1, 1)$ is
$$y - 1 = m(x - 1) \qquad (5.16)$$
Since the above is perpendicular to the line $3x + 4y + 5 =$, whose slope $m' = -\frac{3}{4}$, hence by *Art.* 93, *Cor.* 2,
$$1 + (m + m')\cos\omega + mm' = 0, \text{ where } m' = -\frac{3}{4}$$

5.1. Oblique Coordinates

$$\therefore 1 + \left(m - \frac{3}{4}\right)\cos 120° - \frac{3}{4} \cdot m = 0$$

$$\therefore m = \frac{11}{10}$$

Hence the require equation (5.16) becomes
$$10y - 11x + 1 = 0. \qquad \blacksquare$$

§ Problem 5.1.13. *The coordinates of a point P referred to axes meeting at an angle ω are (h, k); prove that the length of the straight line joining the feet of the perpendiculars from P upon the axes is*
$$\sin\omega\sqrt{h^2 + k^2 + 2hk\cos\omega}. \qquad \Diamond$$

§§ Solution. Let PX and PY be the perpendiculars to the x-axis and y-axis respectively. $O(0,0)$ is the origin.

It is easy to see that
$$OX = h + k\cos\omega, \text{ and}$$
$$OY = k + h\cos\omega.$$

The length of the straight line joining the feet X and Y of the perpendiculars PX and PY from $P(h, k)$ upon the axes is XY, which is computed as follows:

$$XY^2 = OX^2 + OY^2 - 2 \cdot OX \cdot OY \cos\omega$$
$$= (h + k\cos\omega)^2 + (k + h\cos\omega)^2 - 2(h + k\cos\omega)(k + h\cos\omega)\cos\omega$$
$$= (h^2 + k^2 + 2hk\cos\omega)\sin^2\omega$$
$$\therefore XY = \sin\omega\sqrt{h^2 + k^2 + 2hk\cos\omega}. \qquad \blacksquare$$

§ Problem 5.1.14. *From a given point (h, k) perpendiculars are drawn to the axes, whose inclination is ω, and their feet are joined. Prove that the length of the perpendicular drawn from (h, k) upon this line is*
$$\frac{hk\sin^2\omega}{\sqrt{h^2 + k^2 + 2hk\cos\omega}},$$
and that its equation is $hx - ky = h^2 - k^2$. $\qquad \Diamond$

§§ Solution. Let us extend the notations used in the previous exercise. Let PL be the line drawn perpendicular to XY from $P(h, k)$.

We know from the previous exercise that
$$XY = \sin\omega\sqrt{h^2 + k^2 + 2hk\cos\omega}.$$

Area of the $\triangle PXY = \frac{1}{2} \cdot XY \cdot PL$.

Alternatively, $\triangle PXY = \frac{1}{2} \cdot PX \cdot PY \cdot \sin\angle XPY$.

Equating these two expressions of $\triangle PXY$, we get the following:
$$\frac{1}{2} \cdot XY \cdot PL = \frac{1}{2} \cdot PX \cdot PY \cdot \sin\angle XPY$$
$$\therefore \sin\omega\sqrt{h^2 + k^2 + 2hk\cos\omega} \cdot PL = h\sin\omega \cdot k\sin\omega \cdot \sin(\pi - \omega)$$
$$\therefore PL = \frac{hk\sin^2\omega}{\sqrt{h^2 + k^2 + 2hk\cos\omega}}.$$

The equation of XY is
$$\frac{x}{h + k\cos\omega} + \frac{y}{k + h\cos\omega} = 1. \qquad (5.17)$$

Its slope $m' = -\frac{h + k\cos\omega}{k + h\cos\omega}$.

Equation of the line with slope m and passing through the point $P(h, k)$ is
$$y - k = m(x - h). \tag{5.18}$$

If the lines (5.17) and (5.18) are perpendicular to each other then by Art. 93, Cor. 2,
$$1 + (m + m') \cos \omega + mm' = 0$$
$$\therefore 1 + \left(m - \frac{h + k \cos \omega}{k + h \cos \omega}\right) \cos \omega - m \frac{h + k \cos \omega}{k + h \cos \omega} = 0$$
$$\therefore 1 + m \cos \omega = \left(\frac{k + h \cos \omega}{h + k \cos \omega}\right)(\cos \omega + m)$$
$$\therefore h + (mk - h) \cos^2 \omega = mk$$
$$\therefore (mk - h) \sin^2 \omega = 0$$
$$\therefore mk - h = 0$$
$$\therefore m = \frac{h}{k}.$$

Hence the required equation (5.18) becomes
$$ky - k^2 = hx - h^2$$
$$\therefore hx - ky = h^2 - k^2. \quad \blacksquare$$

5.2 Straight lines passing through fixed pts.

§ Problem 5.2.1. *A straight line is such that the algebraic sum of the perpendiculars let fall upon it from any number of fixed points is zero; show that it always passes through a fixed point.* ◊

§§ Solution. Let the equation to the straight line be
$$x \cos \alpha + y \sin \alpha = p \tag{5.19}$$

Let the coordinates the fixed points be
$$(x_1, y_1), (x_2, y_2), \ldots (x_n, y_n)$$

By *Art.* 75, length of the perpendicular drawn upon the line (5.19) from a given fixed point (x_k, y_k) is
$$= \frac{x_k \cos \alpha + y_k \sin \alpha - p}{\sqrt{\cos^2 \alpha + \sin^2 \alpha}} = x_k \cos \alpha + y_k \sin \alpha - p$$

It is given that the sum of the perpendiculars drawn upon the line (5.19) from these points is zero,
$$\therefore (x_1 \cos \alpha + y_1 \sin \alpha - p) + (x_2 \cos \alpha + y_2 \sin \alpha - p) + \ldots$$
$$+ (x_n \cos \alpha + y_n \sin \alpha - p) = 0$$
$$\therefore \sum_{k=1}^{n} x_k \cos \alpha + \sum_{k=1}^{n} y_k \sin \alpha - np = 0$$
$$\therefore \left(\frac{\sum_{k=1}^{n} x_k}{n}\right) \cos \alpha + \left(\frac{\sum_{k=1}^{n} y_k}{n}\right) \sin \alpha = p \tag{5.20}$$

5.2. Straight lines passing through fixed pts.

Comparing the equations (5.19) and (5.20), it is easy to see that the equation (5.19) passes through the fixed point
$$\left(\frac{\sum_{k=1}^{n} x_k}{n}, \frac{\sum_{k=1}^{n} y_k}{n}\right)$$
Alternatively, (5.19) − (5.20) $\implies$
$$\left(x - \frac{\sum_{k=1}^{n} x_k}{n}\right)\cos\alpha + \left(y - \frac{\sum_{k=1}^{n} y_k}{n}\right)\sin\alpha = 0$$
$$\therefore \left(x - \frac{\sum_{k=1}^{n} x_k}{n}\right) + \tan\alpha \left(y - \frac{\sum_{k=1}^{n} y_k}{n}\right) = 0$$

By *Art.* 97, this equation always passes through one fixed point whatever be the value of $\lambda = \tan\alpha$ and that fixed point is the point of intersection of the lines
$$x - \frac{\sum_{k=1}^{n} x_k}{n} = 0$$
$$y - \frac{\sum_{k=1}^{n} y_k}{n} = 0$$
i.e.
$$\left(\frac{\sum_{k=1}^{n} x_k}{n}, \frac{\sum_{k=1}^{n} y_k}{n}\right)$$
∎

§ Problem 5.2.2. *Two fixed straight lines OX and OY are cut by a variable line in the points A and B respectively and P and Q are the feet of the perpendiculars drawn from A and B upon the lines OBY and OAX. Show that, if AB pass through a fixed point, then PQ will also pass through a fixed point.* ◊

§§ Solution. Let OX and OY be the x-axis and y-axis respectively and let the points A and B be $(a, 0)$ and $(0, b)$ respectively, so that $OA = a$ and $OB = b$.

Hence the equation of the line AB is
$$\frac{x}{a} + \frac{y}{b} = 1$$
If the line AB passes through a given point (h, k) then
$$\frac{h}{a} + \frac{k}{b} = 1 \tag{5.21}$$
It is easy to see that
$$OP = OA\cos\omega = a\cos\omega$$
$$OQ = OB\cos\omega = b\cos\omega$$
Hence the equation of the line PQ is
$$\frac{x}{b\cos\omega} + \frac{y}{a\cos\omega} = 1 \tag{5.22}$$

5.2. Straight lines passing through fixed pts.

Comparing the equations (5.21) and (5.22), it is easy to decipher that putting $x = k \cos \omega$ and $y = h \cos \omega$ into (5.22), it becomes identical to (5.21), hence if the equation (5.21) passes through a given fixed point (h, k), then the equation (5.22) passes through a given fixed point $(k \cos \omega, h \cos \omega)$.

Alternatively, subtracting the equations of PQ from AB, we get the following:
$$\frac{x}{a} + \frac{y}{b} = \frac{x}{b \cos \omega} + \frac{y}{a \cos \omega}$$
$$\therefore \left(\frac{x}{a} + \frac{y}{b}\right) + \left(-\frac{1}{\cos \omega}\right)\left(\frac{x}{b} + \frac{y}{a}\right) = 0$$
which passes through one fixed point irrespective of the value of $\lambda = -\dfrac{1}{\cos \omega}$ which is the point of intersection of the following lines
$$\frac{x}{a} + \frac{y}{b} = 0$$
$$\frac{x}{b} + \frac{y}{a} = 0$$
■

§ Problem 5.2.3. *If the equal sides AB and AC of an isosceles triangle be produced to E and F so that $BE.CF = AB^2$, show that the line EF will always pass through a fixed point.* ◇

§§ Solution. Let AB and AC be the x-axis and y-axis respectively such that $AB = AC = a$, and $AE = h$, $AF = k$.

Hence the equation of EF is
$$\frac{x}{h} + \frac{y}{k} = 1 \qquad (5.23)$$

$$\because BE \cdot CF = AB^2$$
$$\therefore (h-a)(k-a) = a^2$$
$$\therefore hk - ha - ka = 0$$
$$\therefore \frac{a}{h} + \frac{a}{k} = 1 \qquad (5.24)$$

Comparing (5.23) and (5.24), it is clear that the line (5.23) always passes through the fixed point (a, a). ■

§ Problem 5.2.4. *If a straight line move so that the sum of the perpendiculars let fall on it from the two fixed points $(3, 4)$ and $(7, 2)$ is equal to three times the perpendicular on it from a third fixed point $(1, 3)$, prove that there is another fixed point through which this line always passes and find its coordinates.* ◇

§§ Solution. Let the equation of the line be
$$x \cos \alpha + y \sin \alpha - p = 0 \qquad (5.25)$$

By Art. 75, length of the perpendicular drawn upon it from a given point (x', y') is
$$= x' \cos \alpha + y' \sin \alpha - p$$

Hence, it is given that
$$(3\cos\alpha + 4\sin\alpha - p) + (7\cos\alpha + 2\sin\alpha - p) = 3(1\cos\alpha + 3\sin\alpha - p)$$
$$\therefore -7\cos\alpha + 3\sin\alpha - p = 0 \qquad (5.26)$$

Comparing (5.25) and (5.26), it is clear that the line (5.25) always passes through the fixed point $(-7, 3)$.

Alternatively, (5.25) − (5.26) $\Longrightarrow$
$$(x + 7) \cos \alpha + (y - 3) \sin \alpha = 0$$

5.2. Straight lines passing through fixed pts.

$$\therefore (x+7) + \tan\alpha(y-3) = 0$$

The above line always passes through a fixed point irrespective of the value of $\lambda = \tan\alpha$, and that fixed point is the point of intersection of the following lines

$$x + 7 = 0$$
$$y - 3 = 0$$

Hence the coordinates of the required point is $(-7, 3)$. ∎

Find the center and radius of the circle which is inscribed in the triangle formed by the straight lines whose equations are
§ Problem 5.2.5. $3x + 4y + 2 = 0$, $3x - 4y + 12 = 0$, and $4x - 3y = 0$. ◊
§§ Solution. Let us assign the given equations to the sides of a triangle ABC, such that the equations represent the sides AB, BC and CA respectively

$$3x + 4y + 2 = 0 \tag{5.27}$$
$$3x - 4y + 12 = 0 \tag{5.28}$$
$$4x - 3y = 0 \tag{5.29}$$

Solving the equations AB: (5.27) and BC: (5.28), we get the coordinates of the angular point B as follows:

(5.27) − (5.28) $\implies$

$$8y - 10 = 0, \therefore y = \frac{5}{4}$$

$$\implies 3x = 4y - 12 = 5 - 12 = -7, \therefore x = -\frac{7}{3}$$

$$\therefore B \equiv \left(-\frac{7}{3}, \frac{5}{4}\right)$$

Solving the equations BC: (5.28) and CA: (5.29), we get the coordinates of the angular point C as follows:

$4 \times (5.28) - 3 \times (5.29) \implies$

$$-7y + 48 = 0, \therefore y = \frac{48}{7}$$

$$\implies 4x = 3y, \therefore x = \frac{36}{7}$$

$$\therefore C \equiv \left(\frac{36}{7}, \frac{48}{7}\right)$$

Solving the equations AB: (5.27) and CA: (5.29), we get the coordinates of the angular point A as follows:

$4 \times (5.27) - 3 \times (5.29) \implies$

$$25y + 8 = 0, \therefore y = -\frac{8}{25}$$

$$\implies 4x = 3y, \therefore x = -\frac{6}{25}$$

$$\therefore A \equiv \left(-\frac{6}{25}, -\frac{8}{25}\right)$$

From the figure, it is easy to infer that the origin lies inside the angle $\angle ABC$ and outside the angle $\angle BCA$.

Hence from the *Art.* 84, the equation of the internal bisector of the angle $\angle ABC$ is given by

$$\frac{3x + 4y + 2}{\sqrt{3^2 + 4^2}} = +\frac{3x - 4y + 12}{\sqrt{3^2 + (-4)^2}}$$

$$\therefore y = \frac{5}{4}$$

5.2. Straight lines passing through fixed pts.

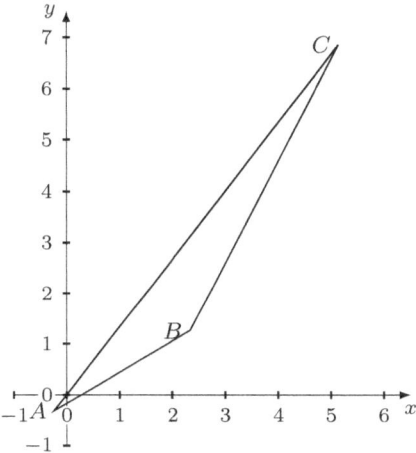

And the equation of the internal bisector of the angle $\angle BCA$ is given by
$$\frac{3x - 4y + 12}{\sqrt{3^2 + (-4)^2}} = -\frac{4x - 3y}{\sqrt{4^2 + (-3)^2}}$$
$$\therefore 7x - 7y + 12 = 0$$

Solving these two equations, we get
$$7x = 7y - 12 = \frac{35}{4} - 12 = -\frac{13}{4}, \therefore x = -\frac{13}{28}.$$

The point of intersection $\left(-\frac{13}{28}, \frac{5}{4}\right)$ of these internal bisectors is the required incenter.

And the radius of incircle is given by the length of the perpendicular from the incenter upon the side AC, i.e.,
$$= \frac{4 \cdot -\frac{13}{28} - 3 \cdot \frac{5}{4}}{\sqrt{4^2 + (-3)^2}} = -\frac{157}{140}.$$

Hence the radius of the incircle is $\frac{157}{140}$.

Alternative Solution :

By *Art.* 98, the coordinates of the incenter is given by
$$\left(\frac{ax_1 + bx_2 + cx_3}{a+b+c}, \frac{ay_1 + by_2 + cy_3}{a+b+c}\right)$$
where
$$A \equiv (x_1, y_1) \equiv \left(-\frac{6}{25}, -\frac{8}{25}\right)$$
$$B \equiv (x_2, y_2) \equiv \left(-\frac{7}{3}, \frac{5}{4}\right)$$
$$C \equiv (x_3, y_3) \equiv \left(\frac{36}{7}, \frac{48}{7}\right)$$

5.2. Straight lines passing through fixed pts.

and
$$a = BC = \sqrt{\left(-\frac{7}{3}-\frac{36}{7}\right)^2 + \left(\frac{5}{4}-\frac{48}{7}\right)^2} = \frac{785}{84}.$$
$$b = AC = \sqrt{\left(-\frac{6}{25}-\frac{36}{7}\right)^2 + \left(-\frac{8}{25}-\frac{48}{7}\right)^2} = \frac{314}{35}.$$
$$c = AB = \sqrt{\left(-\frac{6}{25}+\frac{7}{3}\right)^2 + \left(-\frac{8}{25}-\frac{5}{4}\right)^2} = \frac{157}{60}$$

Putting these values, we get the coordinates of the incenter as $\left(-\frac{13}{28}, \frac{5}{4}\right)$. ∎

§ Problem 5.2.6. $2x + 4y + 3 = 0$, $4x + 3y + 3 = 0$, and $x + 1 = 0$. ◊

§§ Solution. Let us assign the given equations to the sides of a triangle ABC, such that the equations represent the sides AB, BC and CA respectively

$$2x + 4y + 3 = 0 \tag{5.30}$$
$$4x + 3y + 3 = 0 \tag{5.31}$$
$$x + 1 = 0 \tag{5.32}$$

Solving the equations AB: (5.30) and BC: (5.31), we get the coordinates of the angular point B as follows:

$2 \times$ (5.30) $-$ (5.31) $\implies$

$$5y + 3 = 0, \therefore y = -\frac{3}{5}.$$
$$\implies x = -\frac{3}{10}.$$
$$\therefore B \equiv \left(-\frac{3}{10}, -\frac{3}{5}\right)$$

Solving the equations BC: (5.31) and CA: (5.32), we get the coordinates of the angular point C as follows:

$$x = -1, \ y = \frac{1}{3}$$
$$\therefore C \equiv \left(-1, \frac{1}{3}\right)$$

Solving the equations AB: (5.30) and CA: (5.32), we get the coordinates of the angular point A as follows:

$$x = -1, \ y = -\frac{1}{4}$$
$$\therefore A \equiv \left(-1, -\frac{1}{4}\right)$$

From the figure, it is easy to infer that the origin lies inside the angle $\angle BAC$ and outside the angle $\angle BCA$.

Hence from the *Art.* 84, the equation of the internal bisector of the angle $\angle BAC$ is given by

$$\frac{2x + 4y + 3}{\sqrt{2^2 + 4^2}} = +(x+1)$$
$$\therefore x(2 - 2\sqrt{5}) + 4y + (3 - 2\sqrt{5}) = 0$$

And the equation of the internal bisector of the angle $\angle BCA$ is given by

$$\frac{4x + 3y + 3}{\sqrt{4^2 + 3^2}} = -(x+1)$$

5.2. Straight lines passing through fixed pts. 113

$$\therefore 9x + 3y + 8 = 0.$$

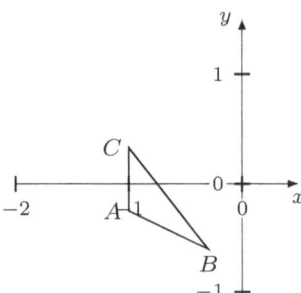

Solving these two equations, we get the point of intersection of these internal bisectors as $\left(\dfrac{-85 - 7\sqrt{5}}{120}, \dfrac{21\sqrt{5} - 65}{120}\right)$, which is the required incenter.

And the radius of incircle is given by the length of the perpendicular from the incenter upon the side AC, i.e.,

$$= \left(-\dfrac{-85 - 7\sqrt{5}}{120} + 1\right) = \dfrac{35 - 7\sqrt{5}}{120}. \qquad \blacksquare$$

§ Problem 5.2.7. $y = 0$, $12x - 5y = 0$, and $3x + 4y - 7 = 0$. $\diamond$

§§ Solution. Let us assign the given equations to the sides of a triangle ABC, such that the equations represent the sides AB, BC and CA respectively

$$y = 0 \tag{5.33}$$
$$12x - 5y = 0 \tag{5.34}$$
$$3x + 4y - 7 = 0 \tag{5.35}$$

Solving the equations AB: (5.33) and BC: (5.34), we get the coordinates of the angular point B as follows:

$$\therefore B \equiv (0, 0)$$

Solving the equations BC: (5.34) and CA: (5.35), we get the coordinates of the angular point C as follows:

$(5.34) - 4 \times (5.35) \implies$

$$-21y + 28 = 0, \therefore y = \dfrac{4}{3}$$
$$\implies x = \dfrac{5}{9}$$
$$\therefore C \equiv \left(\dfrac{5}{9}, \dfrac{4}{3}\right)$$

Solving the equations AB: (5.33) and CA: (5.35), we get the coordinates of the angular point A as follows:

$$\therefore A \equiv \left(\dfrac{7}{3}, 0\right)$$

From the figure and from the *Art.* 84, the equation of the internal bisector of the angle $\angle ABC$ is given by

$$\dfrac{12x - 5y}{13} = +y$$

5.2. Straight lines passing through fixed pts.

$$\therefore x + 3y = \frac{7}{3}$$

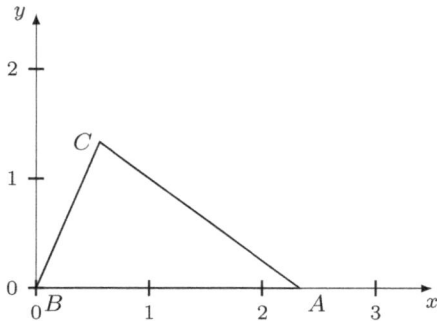

And the equation of the internal bisector of the angle $\angle BAC$ is given by

$$\frac{3x + 4y - 7}{5} = -y$$
$$\therefore 2x = 3y.$$

Solving these two equations, we get the point of intersection of these internal bisectors as $\left(\frac{7}{9}, \frac{14}{27}\right)$, which is the required incenter.

And the radius of incircle is given by the length of the perpendicular from the incenter upon the side $AB : x - axis(y = 0)$, which is $\frac{14}{27}$. ■

§ Problem 5.2.8. *Prove that the coordinates of the center of the circle inscribed in the triangle whose angular points are $(1,2)$, $(2,3)$, and $(3,1)$ are*

$$\frac{8 + \sqrt{10}}{6} \text{ and } \frac{16 - \sqrt{10}}{6}.$$

Find also the coordinates of the centres of the escribed circles. ◊

§§ Solution. Let the angular points be $A \equiv (x_1, y_1) \equiv (1,2)$, $B \equiv (x_2, y_2) \equiv (2,3)$ and $C \equiv (x_3, y_3) \equiv (3,1)$.

Then
$$a = BC = \sqrt{(3-2)^2 + (1-3)^2} = \sqrt{5}.$$
$$b = CA = \sqrt{(3-1)^2 + (1-2)^2} = \sqrt{5}.$$
$$c = AB = \sqrt{(2-1)^2 + (3-2)^2} = \sqrt{2}.$$

Hence the coordinates of the incenter are
$$= \left(\frac{ax_1 + bx_2 + cx_3}{a+b+c}\right) = \left(\frac{\sqrt{5} + 2\sqrt{5} + 3\sqrt{2}}{\sqrt{5} + \sqrt{5} + \sqrt{2}}, \frac{2\sqrt{5} + 3\sqrt{5} + \sqrt{2}}{\sqrt{5} + \sqrt{5} + \sqrt{2}}\right)$$
$$= \left(\frac{8 + \sqrt{10}}{6}, \frac{16 - \sqrt{10}}{6}\right)$$

The coordinates of the center of the escribed circle opposite to A are
$$= \left(\frac{-ax_1 + bx_2 + cx_3}{-a+b+c}\right) = \left(\frac{-\sqrt{5} + 2\sqrt{5} + 3\sqrt{2}}{-\sqrt{5} + \sqrt{5} + \sqrt{2}}, \frac{-2\sqrt{5} + 3\sqrt{5} + \sqrt{2}}{-\sqrt{5} + \sqrt{5} + \sqrt{2}}\right)$$

5.2. Straight lines passing through fixed pts.

$$= \left(\frac{6+\sqrt{10}}{2}, \frac{2+\sqrt{10}}{2} \right)$$

The coordinates of the center of the escribed circle opposite to B are

$$= \left(\frac{ax_1 - bx_2 + cx_3}{a-b+c} \right) = \left(\frac{\sqrt{5} - 2\sqrt{5} + 3\sqrt{2}}{\sqrt{5} - \sqrt{5} + \sqrt{2}}, \frac{2\sqrt{5} - 3\sqrt{5} + \sqrt{2}}{\sqrt{5} - \sqrt{5} + \sqrt{2}} \right)$$

$$= \left(\frac{6-\sqrt{10}}{2}, \frac{2-\sqrt{10}}{2} \right)$$

The coordinates of the center of the escribed circle opposite to C are

$$= \left(\frac{ax_1 + bx_2 - cx_3}{a+b-c} \right) = \left(\frac{\sqrt{5} + 2\sqrt{5} - 3\sqrt{2}}{\sqrt{5} + \sqrt{5} - \sqrt{2}}, \frac{2\sqrt{5} + 3\sqrt{5} - \sqrt{2}}{\sqrt{5} + \sqrt{5} - \sqrt{2}} \right)$$

$$= \left(\frac{8-\sqrt{10}}{6}, \frac{16+\sqrt{10}}{6} \right)$$ ∎

§ Problem 5.2.9. *Find the coordinates of the centers, and the radii, of the four circles which touch the sides of the triangle the coordinates of whose angular points are the points $(6,0)$, $(0,6)$, and $(7,7)$.* ◇

§§ Solution. Let the angular points be $A \equiv (x_1, y_1) \equiv (6,0)$, $B \equiv (x_2, y_2) \equiv (0,6)$ and $C \equiv (x_3, y_3) \equiv (7,7)$.

Then
$$a = BC = \sqrt{7^2 + 1^2} = 5\sqrt{2}.$$
$$b = CA = \sqrt{1^2 + 7^2} = 5\sqrt{2}.$$
$$c = AB = \sqrt{6^2 + 6^2} = 6\sqrt{2}.$$

Hence the coordinates of the incenter are
$$= \left(\frac{ax_1 + bx_2 + cx_3}{a+b+c} \right)$$
$$= \left(\frac{30\sqrt{2} + 42\sqrt{2}}{5\sqrt{2} + 5\sqrt{2} + 6\sqrt{2}}, \frac{30\sqrt{2} + 42\sqrt{2}}{5\sqrt{2} + 5\sqrt{2} + 6\sqrt{2}} \right)$$
$$= \left(\frac{9}{2}, \frac{9}{2} \right).$$

Equation of the line AB passing through the points $A(6,0)$ and $B(0,6)$ is given by
$$\frac{x}{6} + \frac{y}{6} = 1$$
$$\therefore x + y - 6 = 0.$$

The radius of the incircle is the length of the perpendicular from the incenter: $\left(\frac{9}{2}, \frac{9}{2} \right)$ upon the side $AB : x+y-6=0$, i.e.

$$= \frac{\frac{9}{2} + \frac{9}{2} - 6}{\sqrt{1+1}} = \frac{3}{\sqrt{2}} = \frac{3}{2}\sqrt{2}.$$

The coordinates of the center of the escribed circle opposite to A are
$$= \left(\frac{-ax_1 + bx_2 + cx_3}{-a+b+c} \right)$$
$$= \left(\frac{-30\sqrt{2} + 42\sqrt{2}}{-5\sqrt{2} + 5\sqrt{2} + 6\sqrt{2}}, \frac{30\sqrt{2} + 42\sqrt{2}}{-5\sqrt{2} + 5\sqrt{2} + 6\sqrt{2}} \right)$$

$= (2, 12).$

The radius of this circle is
$$= \frac{2+12-6}{\sqrt{1+1}} = \frac{8}{\sqrt{2}} = 4\sqrt{2}.$$

The coordinates of the center of the escribed circle opposite to B are
$$= \left(\frac{ax_1 - bx_2 + cx_3}{a-b+c}\right)$$
$$= \left(\frac{30\sqrt{2} + 42\sqrt{2}}{5\sqrt{2} - 5\sqrt{2} + 6\sqrt{2}}, \frac{-30\sqrt{2} + 42\sqrt{2}}{5\sqrt{2} - 5\sqrt{2} + 6\sqrt{2}}\right)$$
$$= (12, 2).$$

The radius of this circle is
$$= \frac{12 + 2 - 6}{\sqrt{1+1}} = \frac{8}{\sqrt{2}} = 4\sqrt{2}.$$

The coordinates of the center of the escribed circle opposite to C are
$$= \left(\frac{ax_1 + bx_2 - cx_3}{a+b-c}\right)$$
$$= \left(\frac{30\sqrt{2} - 42\sqrt{2}}{5\sqrt{2} + 5\sqrt{2} - 6\sqrt{2}}, \frac{30\sqrt{2} - 42\sqrt{2}}{5\sqrt{2} + 5\sqrt{2} - 6\sqrt{2}}\right)$$
$$= (-3, -3).$$

The radius of this circle is
$$= \frac{-3 - 3 - 6}{\sqrt{1+1}} = -\frac{12}{\sqrt{2}} = 6\sqrt{2}. \qquad \blacksquare$$

§ Problem 5.2.10. *Find the position of the center of the circle circumscribing the triangle whose vertices are the points $(2,3)$, $(3,4)$, and $(6,8)$.* ◊

§§ Solution. Let the angular points be $A(2,3)$, $B(3,4)$ and $C(6,8)$ and let $O(x,y)$ be the coordinates of the circumcenter.

Its radius $= OA = OB = OC$.

$$\therefore OA^2 = OB^2$$
$$\therefore (x-2)^2 + (y-3)^2 = (x-3)^2 + (y-4)^2$$
$$\therefore -4x + 4 - 6y + 9 = -6x + 9 - 8y + 16$$

$$\therefore x + y = 6 \tag{5.36}$$

Similarly,
$$OB^2 = OC^2$$
$$\therefore (x-3)^2 + (y-4)^2 = (x-6)^2 + (y-8)^2$$
$$\therefore -6x + 9 - 8y + 16 = -12x + 36 - 16y + 64$$

$$\therefore 6x + 8y = 75 \tag{5.37}$$

Solving the equations (5.36) and (5.37), we get
$$x = -13\frac{1}{2}, \; y = 19\frac{1}{2}$$

Hence the coordinates of the circum-circle is $\left(-13\frac{1}{2}, 19\frac{1}{2}\right)$.

Alternative Solution :

5.2. Straight lines passing through fixed pts.

The coordinates of the midpoint of the side $AB = \left(\dfrac{2+3}{2}, \dfrac{3+4}{2}\right) = \left(\dfrac{5}{2}, \dfrac{7}{2}\right)$.

Slope of $AB = \dfrac{4-3}{3-2} = 1$.

Hence equation of the line perpendicular to AB and passing through the mid-point of AB is

$$y - \dfrac{7}{2} = -1\left(x - \dfrac{5}{2}\right)$$

$$\therefore x + y = 6 \tag{5.38}$$

The coordinates of the midpoint of the side $BC = \left(\dfrac{3+6}{2}, \dfrac{4+8}{2}\right) = \left(\dfrac{9}{2}, 6\right)$.

Slope of $BC = \dfrac{8-4}{6-3} = \dfrac{4}{3}$.

Hence equation of the line perpendicular to BC and passing through the mid-point of BC is

$$y - 6 = -\dfrac{3}{4}\left(x - \dfrac{9}{2}\right)$$

$$\therefore 6x + 8y = 75 \tag{5.39}$$

It is clear to see that the equation (5.38) is same as (5.36) and (5.39) is same as (5.37).

Solving these equations, we get the coordinates of the circumcircle as $\left(-13\dfrac{1}{2},\ 19\dfrac{1}{2}\right)$. ∎

Find the area of the triangle formed by the straight lines whose equations are

§ Problem 5.2.11. $y = x$, $y = 2x$ and $y = 3x + 4$.   ◊

§§ Solution. Let us assign the given equations to the sides of a triangle ABC, such that the equations represent the sides AB, BC and CA respectively

$$y = x \tag{5.40}$$
$$y = 2x \tag{5.41}$$
$$y = 3x + 4 \tag{5.42}$$

Solving the equations AB: (5.40) and BC: (5.41), we get the coordinates of the angular point B as follows:

$$B \equiv (0, 0)$$

Solving the equations BC: (5.41) and CA: (5.42), we get the coordinates of the angular point C as follows:

$$C \equiv (-4, -8)$$

Solving the equations AB: (5.40) and CA: (5.42), we get the coordinates of the angular point A as follows:

$$A \equiv (-2, -2)$$

By *Art.* 25, the area of the triangle ABC is given by

$$\Delta = \dfrac{1}{2}\begin{vmatrix} x_1 & y_1 & 1 \\ x_2 & y_2 & 1 \\ x_3 & y_3 & 1 \end{vmatrix}$$

5.2. Straight lines passing through fixed pts.   118

As per the *Art.* 27, the area should be a positive quantity and to comply with this, the points A, B and C must be taken in the order in which they would be met by a person starting from A and walking round the triangle in such a manner that the *area of the triangle is always on his left hand.*

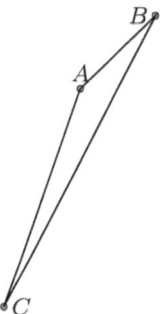

It is clear from the figure that the points should be taken in the order of A, C and B.

$$\therefore \Delta = \frac{1}{2} \begin{vmatrix} [r]-2 & -2 & 1 \\ -4 & -8 & 1 \\ 0 & 0 & 1 \end{vmatrix}$$

$$\therefore \Delta = \frac{1}{2} \{-2(-8-0) + 2(-4-0) + 1(0+0)\}$$
$$= \frac{1}{2}(16-8) = 4.$$

Alternative Solution :

$$BC = \sqrt{4^2 + 8^2} = 4\sqrt{5}.$$

Equation of the side BC is
$$y - 0 = \frac{-8-0}{-4-0}(x-0)$$
$$\therefore y - 2x = 0.$$

Length of the perpendicular drawn from the point $A(-2,-2)$ upon the side BC is
$$= \frac{-2 - 2 \times (-2)}{\sqrt{1^2 + 2^2}} = \frac{2}{\sqrt{5}}.$$
$$\therefore \Delta = \frac{1}{2} \cdot 4\sqrt{5} \cdot \frac{2}{\sqrt{5}} = 4. \qquad \blacksquare$$

§ Problem 5.2.12. $y + x = 0$, $y = x + 6$ *and* $y = 7x + 5$. ◊

§§ Solution. Similar to the previous question, solving the equations we get the angular points as $B(-3, 3)$, $C\left(-\frac{5}{8}, \frac{5}{8}\right)$, $A\left(\frac{1}{6}, \frac{37}{6}\right)$.
$$BC = \sqrt{\left(-\frac{5}{8}+3\right)^2 + \left(\frac{5}{8}-3\right)^2} = \frac{19}{8}\sqrt{2}.$$

5.2. Straight lines passing through fixed pts.

Slope of the side BC is $= \dfrac{\frac{5}{8} - 3}{-\frac{5}{8} + 3} = -1.$

Equation of the side BC is
$$y - 3 = (-1)(x+3)$$
$$\therefore x + y = 0.$$

Length of the perpendicular drawn from the point $A\left(\dfrac{1}{6}, \dfrac{37}{6}\right)$ upon the side BC is

$$= \dfrac{\frac{1}{6} + \frac{37}{6}}{\sqrt{1^2 + 1^2}} = \dfrac{19}{3\sqrt{2}}.$$

$$\therefore \Delta = \dfrac{1}{2} \cdot \dfrac{19}{8}\sqrt{2} \cdot \dfrac{19}{3\sqrt{2}} = \dfrac{361}{48} = 7\dfrac{25}{48}. \blacksquare$$

§ Problem 5.2.13. $2y + x - 5 = 0$, $y + 2x - 7 = 0$ and $x - y + 1 = 0$. ◊

§§ Solution. Solving the given equations, we get the coordinates of the angular points as
$$(3,1),\ (2,3),\ (1,2)$$

$$\therefore \Delta = \dfrac{1}{2}\begin{vmatrix} 3 & 1 & 1 \\ 2 & 3 & 1 \\ 1 & 2 & 1 \end{vmatrix} = \dfrac{3}{2}. \blacksquare$$

§ Problem 5.2.14. $3x - 4y + 4a = 0$, $2x - 3y + 4a = 0$, and $5x - y + a = 0$, proving also that the feet of the perpendiculars from the origin upon them are collinear. ◊

§§ Solution. Solving the given equations, we get the coordinates of the angular points as
$$(0, a),\ \left(\dfrac{a}{13}, \dfrac{18a}{13}\right),\ (4a, 4a)$$

$$\therefore \Delta = \dfrac{1}{2}\begin{vmatrix} 0 & a & 1 \\ \dfrac{a}{13} & \dfrac{18a}{13} & 1 \\ 4a & 4a & 1 \end{vmatrix} = \dfrac{17a^2}{26}.$$

Equations of the perpendiculars on the sides of the triangle from the origin $(0,0)$ are as follows:
$$4x + 3y = 0 \tag{5.43}$$
$$3x + 2y = 0 \tag{5.44}$$
$$x + 5y = 0 \tag{5.45}$$

To find the feet of these perpendiculars on the sides, we have to find the points of intersection of the perpendiculars with the corresponding sides, which are as follows
$$\left(-\dfrac{12a}{25}, \dfrac{16a}{25}\right);\ \left(-\dfrac{8a}{13}, \dfrac{12a}{13}\right);\ \left(-\dfrac{5a}{26}, \dfrac{a}{26}\right).$$

Let us find the area of the triangle formed by joining the feet of these perpendiculars.

$$\therefore \Delta = \dfrac{1}{2}\begin{vmatrix} -\dfrac{12a}{25} & \dfrac{16a}{25} & 1 \\ -\dfrac{8a}{13} & \dfrac{12a}{13} & 1 \\ -\dfrac{5a}{26} & \dfrac{a}{26} & 1 \end{vmatrix} = 0.$$

5.2. Straight lines passing through fixed pts.

Hence these points are collinear.

Alternative Solution :

The coordinates of the feet of the perpendiculars on the sides can be computed as follows.

The equation $x\cos\alpha + y\sin\alpha = p$ represents a straight line where p is the length of the perpendicular drawn upon this line from the origin and α is the angle made by this perpendicular with the x-axis.

It is clear to see that the coordinates of the foot of the perpendicular is $(p\cos\alpha, p\sin\alpha)$.

Let us rewrite the given equations to the sides of the triangle in this form as follows:

$$\frac{3x}{\sqrt{3^2+4^2}} - \frac{4y}{\sqrt{3^2+4^2}} = -\frac{4a}{\sqrt{3^2+4^2}}, \therefore \frac{3x}{5} - \frac{4y}{5} = -\frac{4a}{5}$$

$$\frac{2x}{\sqrt{2^2+3^2}} - \frac{3y}{\sqrt{2^2+3^2}} = -\frac{4a}{\sqrt{2^2+3^2}}, \therefore \frac{2x}{\sqrt{13}} - \frac{3y}{\sqrt{13}} = -\frac{4a}{\sqrt{2^2+3^2}}$$

$$\frac{5x}{\sqrt{5^2+1^2}} - \frac{y}{\sqrt{5^2+1^2}} = -\frac{a}{\sqrt{5^2+1^2}}, \therefore \frac{5x}{\sqrt{26}} - \frac{y}{\sqrt{26}} = -\frac{a}{\sqrt{26}}.$$

Hence the coordinates of the feet of the perpendiculars are as follows:

$$(p\cos\alpha, p\sin\alpha) \equiv \left(-\frac{4a}{5} \cdot \frac{3}{5}, -\frac{4a}{5} \cdot -\frac{4}{5}\right) \equiv \left(-\frac{12a}{25}, \frac{16a}{25}\right)$$

$$(p\cos\alpha, p\sin\alpha) \equiv \left(-\frac{4a}{\sqrt{13}} \cdot \frac{2}{\sqrt{13}}, -\frac{4a}{\sqrt{13}} \cdot -\frac{3}{\sqrt{13}}\right) \equiv \left(-\frac{8a}{13}, \frac{12a}{13}\right)$$

$$(p\cos\alpha, p\sin\alpha) \equiv \left(-\frac{a}{\sqrt{26}} \cdot \frac{5}{\sqrt{26}}, -\frac{a}{\sqrt{26}} \cdot -\frac{1}{\sqrt{26}}\right) \equiv \left(-\frac{5a}{26}, \frac{a}{26}\right) \blacksquare$$

§ Problem 5.2.15. $y = ax - bc$, $y = bx - ca$, and $y = cx - ab$. ◊

§§ Solution. Solving the given equations, we get the coordinates of the angular points as

$$(-c, -c(a+b)); \; (-a, -a(b+c)); \; (-b, -b(a+c))$$

$$\therefore \Delta = \frac{1}{2}\begin{vmatrix} -c & -c(a+b) & 1 \\ -a & -a(b+c) & 1 \\ -b & -b(a+c) & 1 \end{vmatrix} = \frac{1}{2}(a-b)(b-c)(c-a). \quad \blacksquare$$

§ Problem 5.2.16. $y = m_1 x + \dfrac{a}{m_1}$, $y = m_2 x + \dfrac{a}{m_2}$, and $y = m_3 x + \dfrac{a}{m_3}$. ◊

§§ Solution. Solving the given equations, we get the coordinates of the angular points as

$$\left(\frac{a}{m_1 m_2}, \frac{a}{m_1} + \frac{a}{m_2}\right); \; \left(\frac{a}{m_2 m_3}, \frac{a}{m_2} + \frac{a}{m_3}\right); \; \left(\frac{a}{m_3 m_1}, \frac{a}{m_3} + \frac{a}{m_1}\right)$$

$$\therefore \Delta = \frac{1}{2}\begin{vmatrix} \frac{a}{m_1 m_2} & \frac{a}{m_1} + \frac{a}{m_2} & 1 \\ \frac{a}{m_2 m_3} & \frac{a}{m_2} + \frac{a}{m_3} & 1 \\ \frac{a}{m_3 m_1} & \frac{a}{m_3} + \frac{a}{m_1} & 1 \end{vmatrix}$$

$$= \frac{1}{2}a^2 \Sigma \left(\frac{1}{m_1} + \frac{1}{m_2}\right)\left(\frac{1}{m_2 m_3} - \frac{1}{m_3 m_1}\right)$$

$$= \frac{1}{2}a^2 \Sigma \frac{1}{m_3}\left(\frac{1}{m_1^2} - \frac{1}{m_2^2}\right)$$

5.2. Straight lines passing through fixed pts.

$$= \frac{a^2}{2m_1^2 m_2^2 m_3^2} \Sigma m_3(m_1^2 - m_2^2)$$

$$= \frac{a^2}{2m_1^2 m_2^2 m_3^2}(m_1 - m_2)(m_2 - m_3)(m_3 - m_1). \qquad \blacksquare$$

§ Problem 5.2.17. $y = m_1 x + c_1$, $y = m_2 x + c_2$, and the axis of y. ◊

§§ Solution. Equation of the axis of y is $x = 0$.

Solving the given equations, we get the coordinates of the angular points as

$$(0, c_1); \; (0, c_2); \; \left(\frac{c_2 - c_1}{m_1 - m_2}, \frac{m_1 c_2 - m_2 c_1}{m_1 - m_2}\right)$$

$$\therefore \Delta = \frac{1}{2} \begin{vmatrix} 0 & c_1 & 1 \\ 0 & c_2 & 1 \\ \dfrac{c_2 - c_1}{m_1 - m_2} & \dfrac{m_1 c_2 - m_2 c_1}{m_1 - m_2} & 1 \end{vmatrix}$$

$$= \frac{1}{2} \frac{(c_1 - c_2)^2}{m_1 - m_2}. \qquad \blacksquare$$

§ Problem 5.2.18. $y = m_1 x + c_1$, $y = m_2 x + c_2$, and $y = m_3 x + c_3$. ◊

§§ Solution. Utilizing the solution of the previous exercise:

Area of the triangle formed the lines $y = m_1 x + c_1$, $y = m_2 x + c_2$ and the y-axis is

$$\frac{1}{2} \frac{(c_1 - c_2)^2}{m_1 - m_2}.$$

Area of the triangle formed the lines $y = m_2 x + c_2$, $y = m_3 x + c_3$ and the y-axis is

$$\frac{1}{2} \frac{(c_2 - c_3)^2}{m_2 - m_3}.$$

Area of the triangle formed the lines $y = m_1 x + c_1$, $y = m_3 x + c_3$ and the y-axis is

$$\frac{1}{2} \frac{(c_1 - c_3)^2}{m_1 - m_3}.$$

It is easy to see that the area of the triangle formed the lines $y = m_1 x + c_1$, $y = m_2 x + c_2$ and $y = m_3 x + c_3$ is given by subtracting the area of the third triangle from the sum of the other two triangles, i.e.,

$$\Delta = \frac{1}{2} \frac{(c_1 - c_2)^2}{m_1 - m_2} + \frac{1}{2} \frac{(c_2 - c_3)^2}{m_2 - m_3} - \frac{1}{2} \frac{(c_1 - c_3)^2}{m_1 - m_3}$$

$$\therefore \Delta = \frac{1}{2} \left\{ \frac{(c_1 - c_2)^2}{m_1 - m_2} + \frac{(c_2 - c_3)^2}{m_2 - m_3} + \frac{(c_3 - c_1)^2}{m_3 - m_1} \right\}. \qquad \blacksquare$$

§ Problem 5.2.19. *Prove that the area of the triangle formed by the three straight lines*

$$a_1 x + b_1 y + c_1 = 0,$$
$$a_2 x + b_2 y + c_2 = 0, \; and$$
$$a_3 x + b_3 y + c_3 = 0$$

is

$$\frac{1}{2} \left\{ \begin{vmatrix} a_1 & b_1 & c_1 \\ a_2 & b_2 & c_2 \\ a_3 & b_3 & c_3 \end{vmatrix} \right\}^2 \div (a_1 b_2 - a_2 b_1)(a_2 b_3 - a_3 b_2)(a_3 b_1 - a_1 b_3). \qquad ◊$$

5.2. Straight lines passing through fixed pts.

§§ Solution. Let us find the coordinates of the triangle by solving the equations of its sides:

$$a_1 x + b_1 y + c_1 = 0 \tag{5.46}$$
$$a_2 x + b_2 y + c_2 = 0 \tag{5.47}$$
$$a_3 x + b_3 y + c_3 = 0 \tag{5.48}$$

Solving the equations (5.46) and (5.47), we get:

$$\frac{x}{\begin{vmatrix} b_1 & b_2 \\ c_1 & c_2 \end{vmatrix}} = \frac{y}{\begin{vmatrix} c_1 & c_2 \\ a_1 & a_2 \end{vmatrix}} = \frac{1}{\begin{vmatrix} a_1 & a_2 \\ b_1 & b_2 \end{vmatrix}}$$

$$\therefore x = \frac{\begin{vmatrix} b_1 & b_2 \\ c_1 & c_2 \end{vmatrix}}{\begin{vmatrix} a_1 & a_2 \\ b_1 & b_2 \end{vmatrix}}, \quad y = \frac{\begin{vmatrix} c_1 & c_2 \\ a_1 & a_2 \end{vmatrix}}{\begin{vmatrix} a_1 & a_2 \\ b_1 & b_2 \end{vmatrix}}$$

Similarly, the other two coordinates are as follows:

$$\left(\frac{\begin{vmatrix} b_2 & b_3 \\ c_2 & c_3 \end{vmatrix}}{\begin{vmatrix} a_2 & a_3 \\ b_2 & b_3 \end{vmatrix}}, \frac{\begin{vmatrix} c_2 & c_3 \\ a_2 & a_3 \end{vmatrix}}{\begin{vmatrix} a_2 & a_3 \\ b_2 & b_3 \end{vmatrix}} \right); \quad \left(\frac{\begin{vmatrix} b_3 & b_1 \\ c_3 & c_1 \end{vmatrix}}{\begin{vmatrix} a_3 & a_1 \\ b_3 & b_1 \end{vmatrix}}, \frac{\begin{vmatrix} c_3 & c_1 \\ a_3 & a_1 \end{vmatrix}}{\begin{vmatrix} a_3 & a_1 \\ b_3 & b_1 \end{vmatrix}} \right)$$

If we denote these angular points by (x_1, y_1), (x_2, y_2), (x_3, y_3), then the area of the triangle is given by:

$$\therefore \Delta = \frac{1}{2} \begin{vmatrix} x_1 & y_1 & 1 \\ x_2 & y_2 & 1 \\ x_3 & y_3 & 1 \end{vmatrix}$$

$$= \frac{1}{2} \left\{ \begin{vmatrix} a_1 & b_1 & c_1 \\ a_2 & b_2 & c_2 \\ a_3 & b_3 & c_3 \end{vmatrix} \right\}^2 \div (a_1 b_2 - a_2 b_1)(a_2 b_3 - a_3 b_2)(a_3 b_1 - a_1 b_3).$$

Alternative Solution :

Utilizing the solution of the next exercise, we see that the area of the triangle is given by

$$\frac{1}{2} \frac{\{p_1 \sin(\gamma - \beta) + p_2 \sin(\alpha - \gamma) + p_3 \sin(\beta - \alpha)\}^2}{\sin(\gamma - \beta) \sin(\alpha - \gamma) \sin(\beta - \alpha)}$$

Comparing the equations of the sides, it follows that :

$$\cos \alpha = a_1, \ \sin \alpha = b_1, \ -p_1 = c_1$$
$$\cos \beta = a_2, \ \sin \beta = b_2, \ -p_2 = c_2$$
$$\cos \gamma = a_3, \ \sin \gamma = b_3, \ -p_3 = c_3$$

$$\therefore \Delta = \frac{1}{2} \left\{ \begin{vmatrix} a_1 & b_1 & c_1 \\ a_2 & b_2 & c_2 \\ a_3 & b_3 & c_3 \end{vmatrix} \right\}^2 \div (a_1 b_2 - a_2 b_1)(a_2 b_3 - a_3 b_2)(a_3 b_1 - a_1 b_3).$$

∎

§ Problem 5.2.20. *Prove that the area of the triangle formed by the three straight lines*

$$x \cos \alpha + y \sin \alpha - p_1 = 0,$$
$$x \cos \beta + y \sin \beta - p_2 = 0, \text{ and}$$
$$x \cos \gamma + y \sin \gamma - p_3 = 0,$$

5.2. Straight lines passing through fixed pts.

is
$$\frac{1}{2}\frac{\{p_1\sin(\gamma-\beta)+p_2\sin(\alpha-\gamma)+p_3\sin(\beta-\alpha)\}^2}{\sin(\gamma-\beta)\sin(\alpha-\gamma)\sin(\beta-\alpha)}.$$
◇

§§ **Solution.** Solving the given equations of the sides, the coordinates of the angular points are computed as follows:
$$\left(\frac{p_2\sin\alpha-p_1\sin\beta}{\sin(\alpha-\beta)},\ \frac{p_1\cos\beta-p_2\cos\alpha}{\sin(\alpha-\beta)}\right)$$
$$\left(\frac{p_3\sin\beta-p_2\sin\gamma}{\sin(\beta-\gamma)},\ \frac{p_2\cos\gamma-p_3\cos\beta}{\sin(\beta-\gamma)}\right)$$
$$\left(\frac{p_1\sin\gamma-p_3\sin\alpha}{\sin(\gamma-\alpha)},\ \frac{p_3\cos\alpha-p_1\cos\gamma}{\sin(\gamma-\alpha)}\right)$$

Hence by *Art.* 25,
$$\Delta=\frac{1}{2}\Sigma\left\{\frac{p_3\sin\beta-p_2\sin\gamma}{\sin(\beta-\gamma)}\left(\frac{p_3\cos\alpha-p_1\cos\gamma}{\sin(\gamma-\alpha)}-\frac{p_1\cos\beta-p_2\cos\alpha}{\sin(\alpha-\beta)}\right)\right\}$$

$\therefore 2\Delta\cdot\sin(\alpha-\beta)\sin(\beta-\gamma)\sin(\gamma-\alpha)$
$=\Sigma(p_3\sin\beta-p_2\sin\gamma)\{p_3\cos\alpha\sin(\alpha-\beta)+$
$p_2\cos\alpha\sin(\gamma-\alpha)-p_1(\cos\gamma\sin(\alpha-\beta)+\cos\beta\sin(\gamma-\alpha))\}$
$=\Sigma(p_3\sin\beta-p_2\sin\gamma)$
$\{p_3\cos\alpha\sin(\alpha-\beta)+p_2\cos\alpha\sin(\gamma-\alpha)+p_1\cos\alpha\sin(\beta-\gamma)\}$

$\because \Sigma\cdot\cos\alpha\sin(\beta-\gamma)=0$, the above expression becomes
$2\Delta\cdot\sin(\alpha-\beta)\sin(\beta-\gamma)\sin(\gamma-\alpha)$
$=\{p_1\sin(\beta-\gamma)+p_2\sin(\gamma-\alpha)+p_3\sin(\alpha-\beta)\}$
$\Sigma(p_3\sin\beta\cos\alpha-p_2\sin\gamma\cos\alpha)$
$=-\{p_1\sin(\beta-\gamma)+p_2\sin(\gamma-\alpha)+p_3\sin(\alpha-\beta)\}^2.$

$$\therefore \Delta=\frac{1}{2}\frac{\{p_1\sin(\gamma-\beta)+p_2\sin(\alpha-\gamma)+p_3\sin(\beta-\alpha)\}^2}{\sin(\gamma-\beta)\sin(\alpha-\gamma)\sin(\beta-\alpha)}.$$ ■

§ **Problem 5.2.21.** *Prove that the area of the parallelogram contained by the lines*
$$4y-3x-a=0,$$
$$3y-4x+a=0,$$
$$4y-3x-3a=0,\ and$$
$$3y-4x+2a=0$$
is $\dfrac{2}{7}a^2$. ◇

§§ **Solution.** The equations of the sides of the parallelogram are given as
$$4y-3x-a=0 \qquad (5.49)$$
$$3y-4x+a=0 \qquad (5.50)$$
$$4y-3x-3a=0 \qquad (5.51)$$
$$3y-4x+2a=0 \qquad (5.52)$$

It is easy to see that the area of the parallelogram is twice the area of the triangle consisting of any three vertices of it.

Solving the equations (5.49) and (5.50); (5.49) and (5.52); (5.50) and (5.51), coordinates of the three vertices are given by
$$(a,a);\ \left(\frac{11a}{7},\frac{10a}{7}\right);\ \left(\frac{13a}{7},\frac{15a}{7}\right)$$

5.2. Straight lines passing through fixed pts.

Hence, the area of the parallelogram is

$$= 2\Delta = \begin{vmatrix} a & a & 1 \\ \dfrac{11a}{7} & \dfrac{10a}{7} & 1 \\ \dfrac{13a}{7} & \dfrac{15a}{7} & 1 \end{vmatrix} = a^2 \left(-\dfrac{5}{7} + \dfrac{11}{7} \cdot \dfrac{8}{7} - \dfrac{13}{7} \cdot \dfrac{3}{7} \right) = \dfrac{2}{7} a^2.$$

■

§ **Problem 5.2.22.** *Prove that the area of the parallelogram whose sides are the straight lines*

$$a_1 x + b_1 y + c_1 = 0,$$
$$a_1 x + b_1 y + d_1 = 0,$$
$$a_2 x + b_2 y + c_2 = 0, \text{ and}$$
$$a_2 x + b_2 y + d_2 = 0$$

is $\dfrac{(d_1 - c_1)(d_2 - c_2)}{a_1 b_2 - a_2 b_1}$. ◊

§§ **Solution.** The equations of the sides of the parallelogram are

$$a_1 x + b_1 y + c_1 = 0 \tag{5.53}$$
$$a_1 x + b_1 y + d_1 = 0 \tag{5.54}$$
$$a_2 x + b_2 y + c_2 = 0 \tag{5.55}$$
$$a_2 x + b_2 y + d_2 = 0 \tag{5.56}$$

It is clear to see that the equations (5.53) and (5.54) represent one set of parallel sides. If these are inclined at α to the x-axis, then $\tan \alpha = -\dfrac{a_1}{b_1}$.

The intercept of (5.53) on the x-axis is $= -\dfrac{c_1}{a_1}$. The intercept of (5.54) on the x-axis is $= -\dfrac{d_1}{a_1}$.

Hence the length of the intercept on the x-axis by these lines $= l_1 = \dfrac{d_1 - c_1}{a_1}$.

Similarly, the equations (5.55) and (5.56) represent the other set of parallel sides. If these are inclined at β to the x-axis, then $\tan \beta = -\dfrac{a_2}{b_2}$.

The intercept of (5.55) on the x-axis is $= -\dfrac{c_2}{a_2}$. The intercept of (5.56) on the x-axis is $= -\dfrac{d_2}{a_2}$.

Hence the length of the intercept on the x-axis by these lines $= l_2 = \dfrac{d_2 - c_2}{a_2}$.

If the length of the sides of the parallelogram be represented by p and q, then it is easy to see that

$$\dfrac{p}{l_1} = \dfrac{\sin \alpha}{\sin(\beta - \alpha)}; \quad \dfrac{q}{l_2} = \dfrac{\sin \beta}{\sin(\beta - \alpha)}$$

$$\therefore p = \dfrac{(d_1 - c_1) \sin \alpha}{a_1 \sin(\beta - \alpha)}; \quad q = \dfrac{(d_2 - c_2) \sin \beta}{a_2 \sin(\beta - \alpha)}.$$

Hence, the area of the parallelogram is

$$= pq \sin(\beta - \alpha) = \dfrac{(d_1 - c_1)(d_2 - c_2) \sin \alpha \sin \beta}{a_1 a_2 \sin(\beta - \alpha)}$$

5.2. Straight lines passing through fixed pts.

$$= \frac{(d_1 - c_1)(d_2 - c_2)}{\left(\frac{1}{\tan\alpha} - \frac{1}{\tan\beta}\right)a_1 a_2} = \frac{(d_1 - c_1)(d_2 - c_2)}{\left(-\frac{b_1}{a_1} + \frac{b_2}{a_2}\right)a_1 a_2} = \frac{(d_1 - c_1)(d_2 - c_2)}{a_1 b_2 - a_2 b_1}.$$

Alternative Solution :

Slope of the lines (5.53) and (5.54) $= m_1 = -\frac{a_1}{b_1}$.

Slope of the lines (5.55) and (5.56) $= m_2 = -\frac{a_2}{b_2}$.

Hence the angle between the sides of the parallelogram is given by

$$= \theta = \tan^{-1} \frac{m_1 - m_2}{1 + m_1 m_2} = \tan^{-1} \frac{b_1 a_2 - a_1 b_2}{a_1 a_2 + b_1 b_2}$$

$$\because \tan\theta = \frac{\sin\theta}{\cos\theta} = \frac{\sin\theta}{\sqrt{1 - \sin^2\theta}}$$

$$\therefore \sin\theta = \frac{\tan\theta}{\sqrt{1 + \tan^2\theta}} = \frac{b_1 a_2 - a_1 b_2}{\sqrt{(a_1^2 + b_1^2)(a_2^2 + b_2^2)}}$$

Let he length of the sides of the parallelogram be represented by p and q.

Perpendicular distance between the sides (5.53) and (5.54)

$$= \frac{d_1 - c_1}{\sqrt{a_1^2 + b_1^2}} = q\sin\theta.$$

Perpendicular distance between the sides (5.55) and (5.56)

$$= \frac{d_2 - c_2}{\sqrt{a_2^2 + b_2^2}} = p\sin\theta.$$

$$\therefore p = \frac{d_2 - c_2}{\sqrt{a_2^2 + b_2^2}\sin\theta}$$

Hence, the area of the parallelogram is

$$= p\frac{d_1 - c_1}{\sqrt{a_1^2 + b_1^2}} = \frac{d_2 - c_2}{\sqrt{a_2^2 + b_2^2}\sin\theta} \frac{d_1 - c_1}{\sqrt{a_1^2 + b_1^2}}$$

$$= \frac{(d_1 - c_1)(d_2 - c_2)}{\sqrt{(a_1^2 + b_1^2)(a_2^2 + b_2^2)}} \frac{\sqrt{(a_1^2 + b_1^2)(a_2^2 + b_2^2)}}{b_1 a_2 - a_1 b_2}$$

$$= \frac{(d_1 - c_1)(d_2 - c_2)}{a_1 b_2 - a_2 b_1}. \text{(ignoring the negative sign)} \quad \blacksquare$$

§ Problem 5.2.23. *The vertices of a quadrilateral, taken in order, are the points $(0,0)$, $(4,0)$, $(6,7)$, and $(0,3)$; find the coordinates of the point of intersection of the two lines joining the middle points of opposite sides.* ◊

§§ Solution. Coordinates of the mid-point of $(0,0)$ and $(4,0)$ is $(2,0)$.
Coordinates of the mid-point of $(6,7)$ and $(0,3)$ is $(3,5)$.
Coordinates of the mid-point of $(0,0)$ and $(0,3)$ is $\left(0, \frac{3}{2}\right)$.
Coordinates of the mid-point of $(4,0)$ and $(6,7)$ is $\left(5, \frac{7}{2}\right)$.
Coordinates of the mid-point of $(2,0)$ and $(3,5)$ is $\left(\frac{5}{2}, \frac{5}{2}\right)$.
Coordinates of the mid-point of $\left(0, \frac{3}{2}\right)$ and $\left(5, \frac{7}{2}\right)$ is also $\left(\frac{5}{2}, \frac{5}{2}\right)$.

Hence $\left(\dfrac{5}{2}, \dfrac{5}{2}\right)$ is the required point of intersection.

(The lines joining the mid-points of the opposite sides of any quadrilateral bisect each other.) ∎

§ Problem 5.2.24. *The lines*
$$x + y + 1 = 0,$$
$$x - y + 2 = 0,$$
$$4x + 2y + 3 = 0, \text{ and}$$
$$x + 2y - 4 = 0$$
are the equations to the sides of a quadrilateral taken in order ; find the equations to its three diagonals and the equation to the line on which their middle points lie. ◊

§§ Solution. Equations of the sides are

$$AB : x + y + 1 = 0 \tag{5.57}$$
$$BC : x - y + 2 = 0 \tag{5.58}$$
$$CD : 4x + 2y + 3 = 0 \tag{5.59}$$
$$DA : x + 2y - 4 = 0 \tag{5.60}$$

Coordinates of the intersection of AB and BC is $B \equiv \left(-\dfrac{3}{2}, \dfrac{1}{2}\right)$.

Coordinates of the intersection of BC and CD is $C \equiv \left(-\dfrac{7}{6}, \dfrac{5}{6}\right)$.

Coordinates of the intersection of CD and DA is $D \equiv \left(-\dfrac{7}{3}, \dfrac{19}{6}\right)$.

Coordinates of the intersection of DA and AB is $A \equiv (-6, 5)$.

Coordinates of the intersection of AB and CD is $K \equiv \left(-\dfrac{1}{2}, -\dfrac{1}{2}\right)$.

Coordinates of the intersection of BC and DA is $L \equiv (0, 2)$.

Hence equation of the diagonal AC is

$$y - 5 = \dfrac{\dfrac{5}{6} - 5}{-\dfrac{7}{6} + 6}(x + 6)$$

$$\therefore 25x + 29y + 5 = 0.$$

Hence equation of the diagonal BD is

$$y - \dfrac{1}{2} = \dfrac{\dfrac{19}{6} - \dfrac{1}{2}}{-\dfrac{7}{3} + \dfrac{3}{2}}\left(x + \dfrac{3}{2}\right)$$

$$\therefore 32x + 10y + 43 = 0.$$

Hence equation of the diagonal KL is

$$y + \dfrac{1}{2} = \dfrac{2 + \dfrac{1}{2}}{0 + \dfrac{1}{2}}\left(x + \dfrac{1}{2}\right)$$

$$\therefore y = 5x + 2.$$

Coordinates of the mid-point of the diagonal AC is $P \equiv \left(-\dfrac{43}{12}, \dfrac{35}{12}\right)$.

Coordinates of the mid-point of the diagonal BD is $Q \equiv \left(-\dfrac{23}{12}, \dfrac{11}{6}\right)$.

Coordinates of the mid-point of the diagonal KL is $R \equiv \left(-\dfrac{1}{4}, \dfrac{3}{4}\right)$.

5.2. Straight lines passing through fixed pts.

Equation of the line PR is
$$y - \frac{3}{4} = \frac{\frac{11}{6} - \frac{3}{4}}{-\frac{23}{12} + \frac{1}{4}}\left(x + \frac{1}{4}\right)$$
$$\therefore 52x + 80y = 47 \tag{5.61}$$

It is easy to see that the coordinates of the point $Q \equiv \left(-\frac{23}{12}, \frac{11}{6}\right)$ satisfies the equation (5.61), hence the points P, Q, R are collinear, hence the equation (5.61) is the required equation to the line on which the middle points of the three diagonals lie. ∎

§ Problem 5.2.25. *Show that the orthocentre of the triangle formed by the the three straight lines*
$$y = m_1 x + \frac{a}{m_1},$$
$$y = m_2 x + \frac{a}{m_2}, \text{ and}$$
$$y = m_3 x + \frac{a}{m_3}$$

is the point
$$\left\{-a,\ a\left(\frac{1}{m_1} + \frac{1}{m_2} + \frac{1}{m_3} + \frac{1}{m_1 m_2 m_3}\right)\right\}. \quad \diamond$$

§§ Solution. Equations of the sides are given as:
$$AB: y = m_1 x + \frac{a}{m_1} \tag{5.62}$$
$$BC: y = m_2 x + \frac{a}{m_2} \tag{5.63}$$
$$CA: y = m_3 x + \frac{a}{m_3} \tag{5.64}$$

Coordinates of the point of intersection of AB and BC is $B \equiv \left(\frac{a}{m_1 m_2}, \frac{a}{m_1} + \frac{a}{m_2}\right)$.

Coordinates of the point of intersection of BC and CA is $C \equiv \left(\frac{a}{m_2 m_3}, \frac{a}{m_2} + \frac{a}{m_3}\right)$.

Coordinates of the point of intersection of CA and AB is $A \equiv \left(\frac{a}{m_3 m_1}, \frac{a}{m_3} + \frac{a}{m_1}\right)$.

Equation of a line perpendicular to BC: (5.63) and passing through the angular point A is
$$y - \left(\frac{a}{m_3} + \frac{a}{m_1}\right) = -\frac{1}{m_2}\left(x - \frac{a}{m_3 m_1}\right)$$
$$\therefore y + \frac{x}{m_2} = a\left(\frac{1}{m_3} + \frac{1}{m_1} + \frac{1}{m_1 m_2 m_3}\right) \tag{5.65}$$

Similarly, equation of a line perpendicular to CA: (5.64) and passing through the angular point B is
$$y + \frac{x}{m_3} = a\left(\frac{1}{m_1} + \frac{1}{m_2} + \frac{1}{m_1 m_2 m_3}\right) \tag{5.66}$$

And, equation of a line perpendicular to AB: (5.62) and passing through the angular point B is
$$y + \frac{x}{m_1} = a\left(\frac{1}{m_2} + \frac{1}{m_3} + \frac{1}{m_1 m_2 m_3}\right) \tag{5.67}$$

5.2. Straight lines passing through fixed pts.

Solving the equations (5.65) and (5.66), we get
$$x = -a, \; y = a\left(\frac{1}{m_1} + \frac{1}{m_2} + \frac{1}{m_3} + \frac{1}{m_1 m_2 m_3}\right)$$
It is easy to see that this point satisfies the equation (5.67) too, hence this is the point of intersection of the equations (5.65), (5.66) and (5.67), hence this is the orthocenter with the coordinates :
$$\left\{-a, \; a\left(\frac{1}{m_1} + \frac{1}{m_2} + \frac{1}{m_3} + \frac{1}{m_1 m_2 m_3}\right)\right\} \qquad \blacksquare$$

§ Problem 5.2.26. *A and B are two fixed points whose coordinates are $(3,2)$ and $(5,1)$ respectively ; ABP is an equilateral triangle on the side of AB remote from the origin. Find the coordinates of P and the orthocentre of the triangle ABP.* ◊

§§ Solution. Let the coordinates of P be (x, y).

Let us draw the equilateral triangle ABP and draw the lines AM, PN and BK perpendicular to the x-axis. Let us also draw a line BL parallel to the axis of x meeting AM in L. It is clear that $\angle ABP = 60°$.

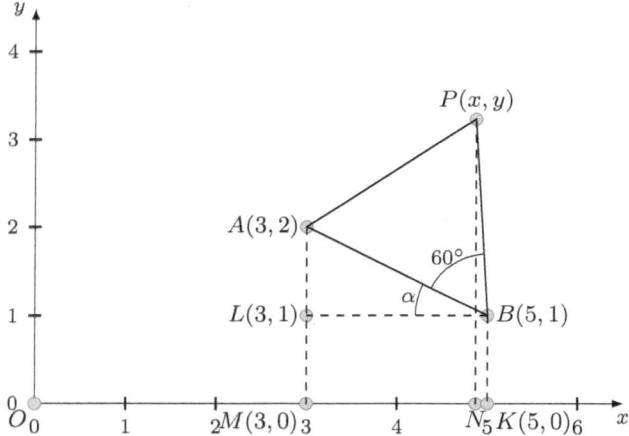

Let $\angle ABL = \alpha$. Then $AB^2 = 1^2 + 2^2$. $\therefore AB = \sqrt{5} = AP = PB$.

It is easy to see from the figure that $\sin\alpha = \dfrac{AL}{AB} = \dfrac{1}{\sqrt{5}}$, $\cos\alpha = \dfrac{BL}{AB} = \dfrac{2}{\sqrt{5}}$,

Since the coordinates of P are (x, y),

$$\therefore x = ON = OK - NK = OK - PB\cdot\cos(\alpha + 60°)$$
$$= 5 - \sqrt{5}\left(\frac{2}{\sqrt{5}}\cdot\frac{1}{2} - \frac{1}{\sqrt{5}}\cdot\frac{\sqrt{3}}{2}\right) = 4 + \frac{\sqrt{3}}{2}.$$
$$\therefore y = PN = BK + PB\cdot\sin(\alpha + 60°)$$
$$= 1 + \sqrt{5}\left\{\frac{1}{\sqrt{5}}\cdot\frac{1}{2} + \frac{2}{\sqrt{5}}\cdot\frac{\sqrt{3}}{2}\right\} = \frac{3}{2} + \sqrt{3}.$$
$$P \equiv (x, y) \equiv \left(4 + \frac{\sqrt{3}}{2}, \; \frac{3}{2} + \sqrt{3}\right).$$

5.2. Straight lines passing through fixed pts.

We know that in an equilateral triangle, the centroid and the orthocentre coincide, hence if h and k are the coordinates of the orthocentre, then by $Ex.2$, page 13 :
$$3h = 3 + 5 + 4 + \frac{\sqrt{3}}{2}, \therefore h = 4 + \frac{\sqrt{3}}{6},$$
$$\text{and } 3k = 2 + 1 + \frac{3}{2} + \sqrt{3}, \therefore k = \frac{3}{2} + \frac{\sqrt{3}}{3}.$$
Hence the coordinates of the orthocenter are =
$$\left(4 + \frac{\sqrt{3}}{6}, \frac{3}{2} + \frac{\sqrt{3}}{3}\right).$$

Alternative Solution :

Since ABP is an equilateral triangle,
$$PA^2 = PB^2 = AB^2$$
$$\therefore (x-3)^2 + (y-2)^2 = (x-5)^2 + (y-1)^2 = 5$$
$$\therefore 4x - 2y - 13 = 0; \; x^2 + y^2 - 6x - 4y + 13 = 5.$$
$$\therefore x^2 + \left(2x - \frac{13}{2}\right)^2 - 6x - 4\left(2x - \frac{13}{2}\right) + 13 = 5$$
$$\therefore 4x^2 - 32x - 61 = 0$$
$$\therefore x = 4 \pm \frac{\sqrt{3}}{2}.$$

Since ABP is an equilateral triangle on the side of AB remote from the origin, hence $x = 4 + \frac{\sqrt{3}}{2}$.
$$\therefore y = \frac{3}{2} + \sqrt{3}.$$
$$P \equiv (x, y) \equiv \left(4 + \frac{\sqrt{3}}{2}, \frac{3}{2} + \sqrt{3}\right).$$
Slope of the side $AB = \frac{2-1}{3-5} = -\frac{1}{2}$.

Equation of the line perpendicular to the side AB and passing through the point P is given by
$$y - \left(\frac{3}{2} + \sqrt{3}\right) = 2\left\{x - \left(4 + \frac{\sqrt{3}}{2}\right)\right\}$$
$$\therefore y = 2x - \frac{13}{2} \tag{5.68}$$

Similarly, equation of the line perpendicular to the side PB and passing through the point A is given by
$$y - 2 = \frac{5\sqrt{3} - 8}{11}(x - 3)$$
$$\therefore 11y - (5\sqrt{3} - 8)x = 46 - 15\sqrt{3}. \tag{5.69}$$

Solving the equations (5.68) and (5.69), we get the following:
$$x = 4 + \frac{\sqrt{3}}{6}, \; y = \frac{3}{2} + \frac{\sqrt{3}}{3}.$$
Hence the coordinates of the orthocenter are =
$$\left(4 + \frac{\sqrt{3}}{6}, \frac{3}{2} + \frac{\sqrt{3}}{3}\right).$$

∎

5.3 Loci

The base $BC(=2a)$ of a triangle ABC is fixed; the axes being BC and a perpendicular to it through its middle pointy find the locus of the vertex A, when

§ Problem 5.3.1. *the difference of the base angles is given $(=\alpha)$.* ◊
§§ Solution. From the hypothesis,
$$\tan\alpha = \frac{\dfrac{y}{a-x} - \dfrac{y}{a+x}}{1 + \dfrac{y^2}{a^2-x^2}}$$
$$\therefore x^2 + 2xy\cot\alpha - y^2 = a^2.\qquad\blacksquare$$

§ Problem 5.3.2. *the product of the tangents of the base angles is given $(=\lambda)$.* ◊
§§ Solution. From the hypothesis,
$$\frac{y}{a-x}\cdot\frac{y}{a+x} = \lambda$$
$$\therefore y^2 + \lambda x^2 = \lambda a^2.\qquad\blacksquare$$

§ Problem 5.3.3. *the tangent of one base angle is m times the tangent of the other.* ◊
§§ Solution. From the hypothesis,
$$\frac{y}{a-x} = m\cdot\frac{y}{a+x}$$
$$\therefore (m+1)x = (m-1)a.\qquad\blacksquare$$

§ Problem 5.3.4. *m times the square of one side added to n times the square of the other side is equal to a constant quantity c^2.* ◊
§§ Solution. From the hypothesis,
$$n\{y^2 + (a+x)^2\} + m\{y^2 + (a-x)^2\} = c^2$$
$$\therefore (m+n)(x^2+y^2+a^2) - 2ax(m-n) = c^2.\qquad\blacksquare$$

From a point P perpendiculars PM and PN are drawn upon two fixed lines which are inclined at an angle ω, and which are taken as the axes of coordinates and meet in O; find the locus of P

§ Problem 5.3.5. *if $OM + ON$ be equal to $2c$.* ◊
§§ Solution. From the hypothesis,
$$(x + y\cos\omega) + (y + x\cos\omega) = 2c$$
$$\therefore x(1+\cos\omega) + y(1+\cos\omega) = 2c$$
$$\therefore x\cdot 2\cos^2\frac{\omega}{2} + y\cdot 2\cos^2\frac{\omega}{2} = 2c$$
$$\therefore x + y = c\cdot\sec^2\frac{\omega}{2}.\qquad\blacksquare$$

§ Problem 5.3.6. *if $OM - ON$ be equal to $2d$.* ◊
§§ Solution. From the hypothesis,
$$(x + y\cos\omega) - (y + x\cos\omega) = 2d$$
$$\therefore x(1-\cos\omega) - y(1-\cos\omega) = 2d$$
$$\therefore x\cdot 2\sin^2\frac{\omega}{2} - y\cdot 2\sin^2\frac{\omega}{2} = 2d$$
$$\therefore x - y = d\cdot\operatorname{cosec}^2\frac{\omega}{2}.\qquad\blacksquare$$

5.3. Loci

§ Problem 5.3.7. *if $PM + PN$ be equal to $2c$.* ◊

§§ Solution. From the hypothesis,
$$x\sin\omega + y\sin\omega = 2c$$
$$\therefore x + y = 2c \cdot \text{cosec}\,\omega.$$ ∎

§ Problem 5.3.8. *if $PM - PN$ be equal to $2c$.* ◊

§§ Solution. From the hypothesis,
$$y\sin\omega - x\sin\omega = 2c$$
$$\therefore y - x = 2c \cdot \text{cosec}\,\omega.$$ ∎

§ Problem 5.3.9. *if MN be equal to $2c$.* ◊

§§ Solution. From the hypothesis,
$$MN = 2c$$
$$\therefore MN^2 = 4c^2$$
$$\triangle MON : OM^2 + ON^2 - 2 \cdot OM \cdot ON \cdot \cos\omega = 4c^2$$
$$\therefore (x + y\cos\omega)^2 + (y + x\cos\omega)^2 - 2(x + y\cos\omega)(y + x\cos\omega)\cos\omega = 4c^2$$
$$\therefore x^2 + y^2 + 2xy\cos\omega = 4c^2 \text{cosec}^2\omega.$$

Alternative Solution :

$$\triangle MPN : PM^2 + PN^2 - 2 \cdot PM \cdot PN \cdot \cos\angle MPN = 4c^2$$
$$\therefore (y\sin\omega)^2 + (x\sin\omega)^2 - 2(y\sin\omega)(x\sin\omega)\cos(180° - \omega) = 4c^2$$
$$\therefore x^2 + y^2 + 2xy\cos\omega = 4c^2 \text{cosec}^2\omega.$$ ∎

§ Problem 5.3.10. *if MN pass through the fixed point (a,b).* ◊

§§ Solution. Let (h, k) be the coordinates of the point P.
The equation of MN is :
$$\frac{x}{h + k\cos\omega} + \frac{y}{k + h\cos\omega} = 1$$
This passes through the point (a, b) :
$$\therefore \frac{a}{h + k\cos\omega} + \frac{b}{k + h\cos\omega} = 1$$
Hence, the locus of the point P is as follows :
$$\frac{a}{x + y\cos\omega} + \frac{b}{y + x\cos\omega} = 1$$
$$\therefore x(a\cos\omega + b) + y(a + b\cos\omega) = (x^2 + y^2)\cos\omega + xy(1 + \cos^2\omega).$$ ∎

§ Problem 5.3.11. *if MN be parallel to the given line $y = mx$.* ◊

§§ Solution. From the previous solution, the slope of MN is
$$-\frac{k + h\cos\omega}{h + k\cos\omega}.$$
From the hypothesis, this is parallel to $y = mx$,
$$\therefore m = -\frac{k + h\cos\omega}{h + k\cos\omega}.$$
Hence, the locus of the point P is as follows :
$$m = -\frac{y + x\cos\omega}{x + y\cos\omega}$$
$$\therefore x(m + \cos\omega) + y(1 + m\cos\omega) = 0.$$ ∎

5.3. Loci

§ Problem 5.3.12. *Two fixed points A and B are taken on the axes such that $OA = a$ and $OB = b$; two variable points A' and B' are taken on the same axes; find the locus of the intersection of AB' and $A'B$*

(1) when $OA' + OB' = OA + OB$, and

(2) when $\dfrac{1}{OA'} - \dfrac{1}{OB'} = \dfrac{1}{OA} - \dfrac{1}{OB}$.

◊

§§ Solution. From the first hypothesis, (1)
$$a' + b' = a + b$$
$$\therefore a - a' = b' - b \tag{5.70}$$
The equation to the straight line AB' is
$$\frac{x}{a} + \frac{y}{b'} = 1 \tag{5.71}$$
The equation to the straight line $A'B$ is
$$\frac{x}{a'} + \frac{y}{b} = 1 \tag{5.72}$$
$(5.71) - (5.72) \Longrightarrow$
$$x\left(\frac{1}{a} - \frac{1}{a'}\right) + y\left(\frac{1}{b'} - \frac{1}{b}\right) = 0$$
$$\therefore \frac{x}{aa'}(a' - a) + \frac{y}{bb'}(b - b') = 0$$
$$\therefore \frac{x}{aa'} + \frac{y}{bb'} = 0. \ (\because \text{from (5.71)}: a - a' = b' - b) \tag{5.73}$$
$(5.71) \times \dfrac{1}{b} + (5.72) \times \dfrac{1}{a} \Longrightarrow$
$$\frac{x}{ab} + \frac{y}{ab} + \frac{x}{aa'} + \frac{y}{bb'} = \frac{1}{a} + \frac{1}{b}$$
$$\therefore \frac{x}{ab} + \frac{y}{ab} = \frac{1}{a} + \frac{1}{b} \ \left(\because \text{from (5.73)}: \frac{x}{aa'} + \frac{y}{bb'} = 0\right)$$
$$\therefore x + y = a + b$$

Alternative Solution :

From (5.71) :
$$b' = \frac{y}{1 - \dfrac{x}{a}}$$
From (5.72) :
$$a' = \frac{x}{1 - \dfrac{y}{b}}$$
Substituting these values in (5.70), we get
$$\frac{x}{1 - \dfrac{y}{b}} + \frac{y}{1 - \dfrac{x}{a}} = a + b$$
$$\therefore bx^2 + (a+b)xy + ay^2 - bx(2a+b) - ay(a+2b) + ab(a+b) = 0$$
$$\therefore (x + y - a - b)(bx + ay - ab) = 0$$
$$\therefore \text{either } x + y - a - b = 0, \text{ or } bx + ay - ab = 0$$
$$\therefore \text{either } x + y = a + b, \text{ or } \frac{x}{a} + \frac{y}{b} = 1$$

The second factor $\dfrac{x}{a} + \dfrac{y}{b} = 1$ represents the straight line AB itself, which is the case when A' coincides with A, B' coincides with B and

5.3. Loci

$A'B'$ coincides with AB, whereas the second factor $x + y = a + b$ represents the required locus.

From the second hypothesis, (2)
$$\frac{1}{OA'} - \frac{1}{OB'} = \frac{1}{OA} - \frac{1}{OB}$$
$$\therefore \frac{1}{a'} - \frac{1}{b'} = \frac{1}{a} - \frac{1}{b}$$
$$\therefore \frac{1}{a} - \frac{1}{a'} = \frac{1}{b} - \frac{1}{b'} \tag{5.74}$$

$(5.71) - (5.72) \implies$
$$x\left(\frac{1}{a} - \frac{1}{a'}\right) = y\left(\frac{1}{b} - \frac{1}{b'}\right)$$
$$\therefore x = y. \ \left(\because \text{from (5.74)}: \frac{1}{a} - \frac{1}{a'} = \frac{1}{b} - \frac{1}{b'}\right) \qquad \blacksquare$$

§ Problem 5.3.13. *Through a fixed point P are drawn any two straight lines to cut one fixed straight line OX in A and B and another fixed straight line OY in C and D ; prove that the locus of the intersection of the straight lines AC and BD is a straight line passing through O.* ◊

§§ Solution. Let (h, k) be the coordinates of the point P and let $OA = a$, $OB = b$, $OC = c$ and $OD = d$.

Equation of AD is
$$\therefore \frac{x}{a} + \frac{y}{d} = 1$$
Since this passes through $P(h, k)$,
$$\therefore \frac{h}{a} + \frac{k}{d} = 1 \tag{5.75}$$
Equation of BC is
$$\therefore \frac{x}{b} + \frac{y}{c} = 1$$
Since this passes through $P(h, k)$,
$$\therefore \frac{h}{b} + \frac{k}{c} = 1 \tag{5.76}$$
$(5.75) - (5.76) \implies$
$$\therefore h\left(\frac{1}{a} - \frac{1}{b}\right) + k\left(\frac{1}{d} - \frac{1}{c}\right) = 0 \tag{5.77}$$
Equation of AC is
$$\therefore \frac{x}{a} + \frac{y}{c} = 1 \tag{5.78}$$
Equation of BD is
$$\therefore \frac{x}{b} + \frac{y}{d} = 1 \tag{5.79}$$
$(5.78) - (5.79) \implies$
$$\therefore x\left(\frac{1}{a} - \frac{1}{b}\right) + y\left(\frac{1}{c} - \frac{1}{d}\right) = 0 \tag{5.80}$$
From (5.77) and (5.80), it is easy to infer that
$$\frac{h}{k} = -\frac{x}{y}$$
$$\therefore y = -\frac{k}{h}x.$$
It is a straight line passing through the origin O. $\qquad \blacksquare$

5.3. Loci   134

§ Problem 5.3.14. *OX and OY are two straight lines at right angles to one another; on OY is taken a fixed point A and on OX any point B; on AB an equilateral triangle is described, its vertex C being on the side of AB away from O. Show that the locus of C is a straight line.* ◊

§§ Solution. Let the coordinates of C be (x, y), $OA = a$, $\angle ABO = \theta$. Let CM be perpendicular to the x-axis and AN be parallel to the x-axis.

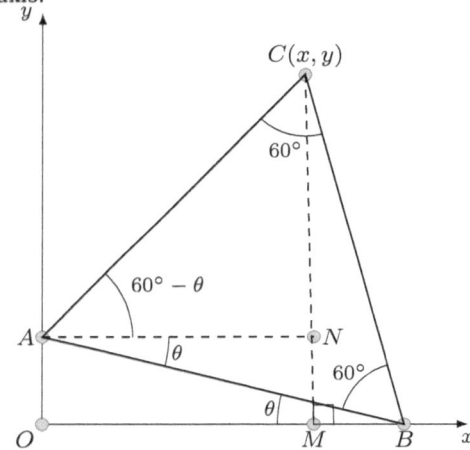

Since ABC is an equilateral triangle, hence $AB = OA \operatorname{cosec} \theta = a \operatorname{cosec} \theta = BC = CA$.

It is clear from the figure, that
$$x = OM = AN = AC\cos(60° - \theta) = a \operatorname{cosec}\theta \cos(60° - \theta)$$
$$\therefore x = a \operatorname{cosec}\theta(\cos 60° \cos\theta + \sin 60° \sin\theta)$$

$$\therefore x = \frac{a}{2}(\sqrt{3} + \cot\theta). \tag{5.81}$$

and, $y = CM = BC \sin \angle MBC = a \operatorname{cosec}\theta \sin(60° + \theta)$

$$\therefore y = \frac{a}{2}(\sqrt{3}\cot\theta + 1). \tag{5.82}$$

$(5.81) \times \sqrt{3} - (5.81) \implies$
$$\sqrt{3}x - y = \frac{3a}{2} - \frac{a}{2}$$
$$\therefore \sqrt{3}x - y = a.$$

It is easy to infer that this is a straight line.

Alternative Solution :

Let us redraw the picture.
It is clear from the figure, that
$$x = OM = OB + BM = AB\cos\theta + BC\cos\angle CBM$$
$$\therefore x = a\operatorname{cosec}\theta\cos\theta + a\operatorname{cosec}\theta\cos(120° - \theta)$$
$$\therefore x = a\cot\theta - \frac{1}{2}a\cot\theta - a \cdot \frac{\sqrt{3}}{2}$$

5.3. Loci

$$\therefore x = \frac{a}{2}(\cot\theta - \sqrt{3}). \tag{5.83}$$

$$y = CM = BC\sin(120° - \theta) = a\,\text{cosec}\,\theta\sin(120° - \theta)$$

$$\therefore y = a\cot\theta \cdot \frac{\sqrt{3}}{2} + \frac{a}{2}$$

$$\therefore y = \frac{a}{2}(\sqrt{3}\cot\theta + 1). \tag{5.84}$$

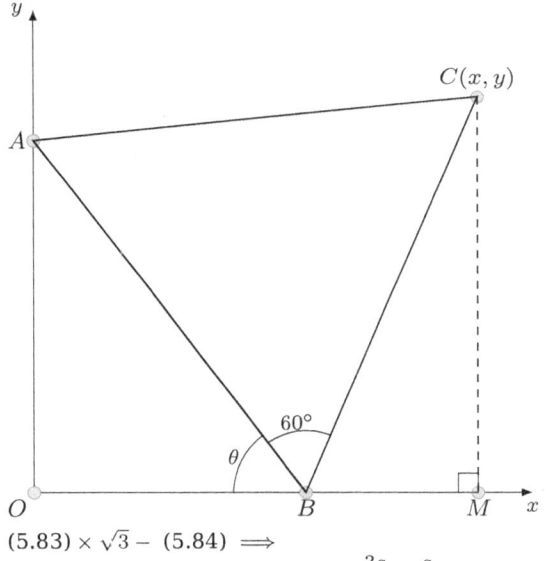

(5.83) × √3 − (5.84) ⟹

$$\sqrt{3}x - y = -\frac{3a}{2} - \frac{a}{2}$$

$$\therefore \sqrt{3}x - y = -2a.$$

It is easy to infer that this is a straight line. ∎

§ Problem 5.3.15. *If a straight line passes through a fixed point, find the locus of the middle point of the portion of it which is intercepted between two given straight lines.* ◊

§§ Solution. Let a and b be the intercepts on the x-axis and y-axis respectively.

The equation of the line is
$$\frac{x}{a} + \frac{y}{b} = 1$$

Let (h, k) be the fixed point through which this line passes,
$$\therefore \frac{h}{a} + \frac{k}{b} = 1$$

Let (x', y') be the mid-point of the intercepted line.

$$\therefore x' = \frac{0+a}{2},\ y' = \frac{b+0}{2}$$
$$\therefore a = 2x',\ b = 2y'$$

5.3. Loci

Hence
$$\frac{h}{2x'} + \frac{k}{2y'} = 1$$
Hence the required locus is
$$\frac{h}{2x} + \frac{k}{2y} = 1$$
∎

§ Problem 5.3.16. *A and B are two fixed points; if PA and PB intersect a constant distance 2c from a given straight line, find the locus of P.* ◊

§§ Solution. For the sake of simplicity, let us take the x-axis as the fixed straight line, the fixed points as $A(0,a)$ and $B(0,b)$.

Let (h,k) be the coordinates of P.

The the equation of AP is
$$y - a = \frac{k-a}{h-0}(x-0) = \frac{k-a}{h}x$$
Putting $y = 0$, we get its intercept on the x-axis as
$$\frac{ah}{a-k} \qquad (5.85)$$
The the equation of AB is
$$y - b = \frac{k-b}{h-0}(x-0) = \frac{k-b}{h}x$$
Putting $y = 0$, we get its intercept on the x-axis as
$$\frac{bh}{b-k} \qquad (5.86)$$
From the given hypothesis, the difference of (5.85) − (5.86) = 2c
$$\therefore \frac{ah}{a-k} - \frac{bh}{b-k} = 2c$$
$$\therefore ah(b-k) - bh(a-k) = 2c(a-k)(b-k)$$
$$\therefore hk(b-a) = 2c(k-a)(k-b)$$
$$\therefore 2c(k-a)(k-b) + (a-b)hk = 0$$
Hence, the required locus of P is given by
$$\therefore 2c(y-a)(y-b) + (a-b)xy = 0$$
∎

§ Problem 5.3.17. *Through a fixed point O are drawn two straight lines at right angles to meet two fixed straight lines, which are also at right angles, in the points P and Q. Show that the locus of the foot of the perpendicular from O on PQ is a straight line.* ◊

§§ Solution. Let $O(0,0)$ be the origin and the lines (at right angles) passing through O be the x-axis and y-axis respectively. Let the other set of straight lines be $O'P$ and $O'Q$ with $O'(h,k)$. For the sake of simplicity, let us assume these sets to be parallel to eaach other.

Let us denote the equation of PQ be
$$x\cos\alpha + y\sin\alpha = p$$
$$\therefore \frac{x\cos\alpha + y\sin\alpha}{p} = 1 \qquad (5.87)$$
Equation of $O'P$ is $x - h = 0$ and of $O'Q$ is $y - k = 0$.

Hence, by *Art.* 106 and 107, the following equation represents lines $O'P$ and $O'Q$:
$$(x-h)(y-k) = 0$$
$$\therefore xy - hy - kx + hk = 0 \qquad (5.88)$$

5.3. Loci

By Art. 122, the equation to the two straight lines joining the origin to the common points of (5.87) and (5.88) is given by
$$xy - (hy + kx)\frac{x\cos\alpha + y\sin\alpha}{p} + hk\left(\frac{x\cos\alpha + y\sin\alpha}{p}\right)^2 = 0$$
$$\therefore p^2 xy - p(hy + kx)(x\cos\alpha + y\sin\alpha) + hk(x\cos\alpha + y\sin\alpha)^2 = 0$$
$$\therefore x^2(-p\cos\alpha k + hk\cos^2\alpha) + xy(2hk\cos\alpha\sin\alpha - hp\cos\alpha - kp\sin\alpha) +$$
$$y^2(-p\sin\alpha h + hk\sin^2\alpha) = 0 \quad (5.89)$$

Since these are at right angles, hence by Art. 110, sum of the coefficients of x^2 and y^2 is 0, i.e.
$$(-p\cos\alpha k + hk\cos^2\alpha) + (-p\sin\alpha h + hk\sin^2\alpha) = 0$$
$$\therefore -p\cos\alpha k - p\sin\alpha h + hk = 0.$$

We know that the coordinates of the foot of the perpendicular from O on PQ is $(p\cos\alpha, p\sin\alpha)$, hence the required locus is given by
$$-xk - yh + hk = 0$$
$$\therefore kx + hy = hk$$
$$\therefore \frac{x}{h} + \frac{y}{k} = 1.$$

This is a straight line. ■

§ Problem 5.3.18. *Find the locus of a point at which two given portions of the same straight line subtend equal angles.* ◊

§§ Solution. Let the straight line be the x-axis with $O(0,0)$ as origin. Let AB and CD be the two given portions of this line such that $OA = a$, $OB = b$, $OC = c$ and $OD = d$.

Let $P(x,y)$ be any point on the locus.
From the hypothesis,
$$\angle APB = \angle CPD$$
$$\therefore \tan\angle APB = \tan\angle CPD$$
$$\therefore \tan(\angle OPB - \angle OPA) = \tan(\angle OPD - \angle OPC)$$
$$\therefore \frac{\frac{y}{x-b} - \frac{y}{x-a}}{1 + \frac{y^2}{(x-b)(x-a)}} = \frac{\frac{y}{x-d} - \frac{y}{x-c}}{1 + \frac{y^2}{(x-d)(x-c)}}$$
$$\therefore \frac{y(b-a)}{(x-a)(x-b) + y^2} = \frac{y(d-c)}{(x-c)(x-d) + y^2}$$
$$\therefore \frac{a-b}{(x-a)(x-b) + y^2} = \frac{c-d}{(x-c)(x-d) + y^2}$$

This is the required locus. ■

§ Problem 5.3.19. *Find the locus of a point which moves so that the difference of its distances from two fixed straight lines at right angles is equal to its distance from a fixed straight line.* ◊

§§ Solution. Let the two fixed lines at right angles be the x-axis and y-axis respectively and let the equation of the other fixed line be
$$x\cos\alpha + y\sin\alpha - p = 0 \quad (5.90)$$

Let $P(h,k)$ be the moving point on the locus. Then its distance from (5.90) is
$$h\cos\alpha + k\sin\alpha - p.$$

5.3. Loci

From the hypothesis,
$$h - k = h\cos\alpha + k\sin\alpha - p$$
$$\therefore h(1 - \cos\alpha) - k(1 + \sin\alpha) + p = 0.$$
Hence the required locus is
$$x(1 - \cos\alpha) - y(1 + \sin\alpha) + p = 0.$$
This is a straight line. ∎

§ **Problem 5.3.20.** *A straight line AB, whose length is c, slides between two given oblique axes which meet at O ; find the locus of the orthocentre of the triangle OAB.* ◊

§§ **Solution.** Let $OA = a$, $OB = b$, $\angle AOB = \omega$. Hence $A \equiv (a, 0)$, $B \equiv (b, 0)$.

By *Art.* 91, the equation of the line inclined at an angle θ to the x-axis and with y-intercept being c is given by
$$y = x\frac{\sin\theta}{\sin(\omega - \theta)} + c \tag{5.91}$$

Putting $\theta = 90°$ in (5.91), we get the equation of line perpendicular to the x-axis as
$$y = x\frac{\sin 90°}{\sin(\omega - 90°)} + c$$
$$y = -\frac{x}{\cos\omega} + c \tag{5.92}$$

Hence, equation of a line passing through $B(0, b)$ and perpendicular to OA is also represented by (5.92).

Putting $x = 0$, $y = b$, this becomes $b = 0 + c$, $\therefore c = b$.
$$\therefore y = -\frac{x}{\cos\omega} + b$$
$$\therefore x + y\cos\omega = b\cos\omega \tag{5.93}$$

Putting $\theta = 90° + \omega$ in (5.91), we get the equation of line perpendicular to the y-axis as
$$y = x\frac{\sin 90° + \omega}{\sin(\omega - 90° - \omega)} + c$$
$$y = -x\cos\omega + c \tag{5.94}$$

Hence, equation of a line passing through $A(a, 0)$ and perpendicular to OB is also represented by (5.93).

Putting $x = a$, $y = 0$, this becomes $0 = -a\cos\omega + c$, $\therefore c = a\cos\omega$.
$$\therefore y = -x\cos\omega + a\cos\omega$$
$$\therefore x\cos\omega + y = a\cos\omega \tag{5.95}$$

Also, from the hypothesis
$$AB = c$$
$$\therefore OA^2 + OB^2 - 2\cdot OA\cdot OB\cos\omega = c^2$$
$$\therefore a^2 + b^2 - 2ab\cos\omega = c^2 \tag{5.96}$$

Putting the values of a from (5.95) and b from (5.93) in (5.96), we get the following
$$(x + y\sec\omega)^2 + (x\sec\omega + y)^2 - 2(x + y\sec\omega)(x\sec\omega + y)\cos\omega = c^2$$
$$x^2 + y^2 + 2xy\cos\omega = c^2\cot^2\omega.$$

This is a circle with O as its centre. This is the locus of the orthocentre of the triangle OAB. ∎

5.3. Loci

§ Problem 5.3.21. *Having given the bases and the sum of the areas of a number of triangles which have a common vertex, show that the locus of this vertex is a straight line.* ◊

§§ Solution. Let $A(h,k)$ be the common vertex. Let l_i be the length of the ith base. Let the equation of the ith base be:
$$x \cos \alpha_i + y \sin \alpha_i = p_i.$$
Length of the perpendicular from $A(h,k)$ upon the ith base is
$$h \cos \alpha_i + k \sin \alpha_i - p_i.$$
Area of the $i^{th} \Delta$ is given by
$$\frac{1}{2} \cdot l_i \cdot (h \cos \alpha_i + k \sin \alpha_i - p_i)$$
From the hypothesis, sum of the areas of the triangles is constant, say c.
$$\therefore \Sigma \frac{1}{2} \cdot l_i \cdot (h \cos \alpha_i + k \sin \alpha_i - p_i) = c$$
$$\therefore h \Sigma l_i \cos \alpha_i + k \Sigma l_i \sin \alpha_i - \Sigma l_i p_i = 2c$$
Hence the locus of the common vertex is
$$\therefore x \Sigma l_i \cos \alpha_i + y \Sigma l_i \sin \alpha_i - \Sigma l_i p_i = 2c$$
This is a straight line. ∎

§ Problem 5.3.22. *Through a given point O a straight line is drawn to cut two given straight lines in R and S ; find the locus of a point P on this variable straight line, which is such that*

(1) $2OP = OR + OS$, and

(2) $OP^2 = OR.OS$. ◊

§§ Solution. By *Art. 105*, let us take any two fixed straight lines, at right angles and passing through O, as the axes and let the equation to the two given fixed straight lines be
$$A_1 x + B_1 y + C_1 = 0,$$
$$A_2 x + B_2 y + C_2 = 0.$$
and, transforming to polar coordinates these equations are
$$\frac{1}{r} = -\frac{A_1 \cos \theta + B_1 \sin \theta}{C_1}, \text{ and}$$
$$\frac{1}{r} = -\frac{A_2 \cos \theta + B_2 \sin \theta}{C_2}$$
If the angle XOR be θ, then the values of $\dfrac{1}{OR}$ and $\dfrac{1}{OS}$ are therefore
$$-\frac{A_1 \cos \theta + B_1 \sin \theta}{C_1}, \text{ and}$$
$$-\frac{A_2 \cos \theta + B_2 \sin \theta}{C_2}$$

(1)
$$2OP = OR + OS$$
$$\therefore 2r = -\frac{C_1}{A_1 \cos \theta + B_1 \sin \theta} - \frac{C_2}{A_2 \cos \theta + B_2 \sin \theta}$$

The equation to the locus of P is therefore, on again transforming to Cartesian coordinates,
$$C_1(A_2x + B_2y) + C_2(A_1x + B_1y) + 2(A_1x + B_1y)(A_2x + B_2y) = 0 \quad (2)$$

$$OP^2 = OR \cdot OS$$
$$\therefore r^2 = -\frac{C_1}{A_1\cos\theta + B_1\sin\theta} \cdot -\frac{C_2}{A_2\cos\theta + B_2\sin\theta}$$

The equation to the locus of P is therefore, on again transforming to Cartesian coordinates,
$$(A_1x + B_1y)(A_2x + B_2y) = C_1C_2.$$ ∎

§ Problem 5.3.23. *Given n straight lines and a fixed point O; through O is drawn a straight line meeting these lines in the points R_1, R_2, R_3, ... R_n, and on it is taken a point R such that*
$$\frac{n}{OR} = \frac{1}{OR_1} + \frac{1}{OR_2} + \frac{1}{OR_3} + \ldots + \frac{1}{OR_n};$$
show that the locus of R is a straight line. ◊

§§ Solution. By *Art.* 105, let us take any two fixed straight lines, at right angles and passing through O, as the axes and let the equation to the ith given fixed straight line be
$$A_ix + B_iy + C_i = 0$$
and, transforming to polar coordinates this equation is
$$\frac{1}{r} = -\frac{A_i\cos\theta + B_i\sin\theta}{C_i}, \text{ and}$$
If the angle XOR be θ, then the value of $\frac{1}{OR_i}$ is therefore
$$-\frac{A_i\cos\theta + B_i\sin\theta}{C_i}$$
From the given hypothesis,
$$\frac{n}{OR} = \sum_{i=1}^{n} -\frac{A_i\cos\theta + B_i\sin\theta}{C_i}$$
$$\frac{n}{r} = -\cos\theta \sum_{i=1}^{n} \frac{A_i}{C_i} - \sin\theta \sum_{i=1}^{n} \frac{B_i}{C_i}$$
The equation to the locus of R is therefore, on again transforming to Cartesian coordinates,
$$x\sum_{i=1}^{n}\frac{A_i}{C_i} + y\sum_{i=1}^{n}\frac{B_i}{C_i} + n = 0$$
This is a straight line. ∎

§ Problem 5.3.24. *A variable straight line cuts off from n given concurrent straight lines intercepts the sum of the reciprocals of which is constant. Show that it always passes through a fixed point.* ◊

§§ Solution. Let origin be the point of concurrence and α_i be the inclination of the ith concurrent line.

Let the equation of the variable line be
$$\frac{1}{r} = a\cos\theta + b\sin\theta$$

5.3. Loci

From the given hypothesis,
$$\sum_{i=1}^{n}(a\cos\alpha_i + b\sin\alpha_i) = c \text{ (a constant)}$$
$$\therefore a\sum_{i=1}^{n}\frac{\cos\alpha_i}{c} + b\sum_{i=1}^{n}\frac{\sin\alpha_i}{c} = 1$$

It is easy to see that this is the condition that the straight line $ax + by = 1$ passes through a fixed point
$$\left(\sum_{i=1}^{n}\frac{\cos\alpha_i}{c}, \sum_{i=1}^{n}\frac{\sin\alpha_i}{c}\right).$$
∎

§ Problem 5.3.25. *If a triangle ABC remain always similar to a given triangle, and if the point A be fixed and the point B always move along a given straight line, find the locus of the point C.* ◊

§§ Solution. Let the fixed point A be the origin. Let (r, θ) be the coordinates of the vertex C. Let the perpendicular from A to locus of B (length c) be the initial line.

From the given hypothesis, $\triangle ABC$ is always similar to the given triangle.
$$\therefore \frac{AB}{AC} = k\text{(a constant)}$$
$$\therefore AB = kr$$
$$\therefore c = AB \cdot \cos(\theta - A) = kr\cos(\theta - A)$$

This is the required locus which is a straight line. ∎

§ Problem 5.3.26. *A right-angled triangle ABC, having C a right angle, is of given magnitude, and the angular points A and B slide along two given perpendicular axes; show that the locus of C is the pair of straight lines whose equations are*
$$y = \pm\frac{b}{c}x.$$
◊

§§ Solution. Let $A(h, 0)$ slide along the x-axis, $B(0, k)$ slide along the y-axis. Let C be (x, y).

$$\therefore BC^2 = (x-0)^2 + (y-k)^2$$
$$\therefore (y-k)^2 + x^2 = a^2 \tag{5.97}$$

and
$$\therefore AC^2 = (x-h)^2 + (y-0)^2$$
$$\therefore (x-h)^2 + y^2 = b^2 \tag{5.98}$$

$$\because AC \perp BC$$
$$\therefore \frac{y-0}{x-h} \cdot \frac{y-k}{x-0} = -1$$
$$\therefore \frac{y^2}{x^2} = \frac{(x-h)^2}{(y-k)^2} = \frac{b^2 - y^2}{a^2 - x^2}, \text{ (From (5.98) and (5.97))}$$
$$\therefore \frac{y^2}{x^2} = \frac{b^2 - y^2}{a^2 - x^2} = \frac{y^2 + (b^2 - y^2)}{x^2 + (a^2 - x^2)} \text{ (Using Componendo and Dividendo)}$$

$$\therefore \frac{y^2}{x^2} = \frac{b^2}{a^2}$$
$$\therefore y = \pm\frac{b}{a}x.$$

Alternative Solution :

Let A slide along the x-axis, B slide along the y-axis and O be the origin. Let C be (x, y).

In the quadrilateral $OACB$:
$$\angle ACB = 90° = BOA$$
$$\therefore \angle ACB + \angle BOA = 180°, \text{ and}$$
$$\angle OAC + CBO = 360° - (\angle ACB + \angle BOA) = 180°.$$

We know that a convex quadrilateral is cyclic if and only if its opposite angles are supplementary, hence $OACB$ is a cyclic quadrilateral.

Another necessary and sufficient condition for a convex quadrilateral to be cyclic is that an angle between a side and a diagonal is equal to the angle between the opposite side and the other diagonal, hence for the cyclic quadrilateral $OACB$, the following holds true :
$$\angle COA = \angle CBA$$
$$\therefore \tan \angle COA = \tan \angle CBA$$
$$\therefore \frac{y}{x} = \frac{AC}{BC} = \frac{b}{a}$$
$$\therefore y = \frac{b}{a}x$$

Similarly, if B lies on the negative side of the y-axis, $y = -\frac{b}{a}x$
$$\therefore y = \pm\frac{b}{a}x.$$ ∎

§ Problem 5.3.27. *Two given straight lines meet in O, and through a given point P is drawn a straight line to meet them in Q and R; if the parallelogram $OQSR$ be completed find the equation to the locus of S.* ◊

§§ Solution. Let Q be $(a, 0)$ and R be $(0, b)$. Let the fixed point P be (h, k).

Then the equation of QR is
$$\frac{x}{a} + \frac{y}{b} = 1$$
Since the point $P(h, k)$ lies on the line QR,
$$\therefore \frac{h}{a} + \frac{k}{b} = 1$$

Since $OQSR$ is a parallelogram, hence the coordinates of S is (a, b). Hence the locus of S is given by
$$\therefore \frac{h}{x} + \frac{k}{y} = 1$$ ∎

§ Problem 5.3.28. *Through a given point O is drawn a straight line to meet two given parallel straight lines in P and Q ; through P and Q are drawn straight lines in given directions to meet in R ; prove that the locus of R is a straight line.* ◊

5.3. Loci

§§ Solution. Let O be the origin. Let the two given parallel lines be perpendicular to the x-axis and their distances from the x-axis be a and b respectively.

Let $\angle POX = \theta$, then it is easy to see that the coordinates of P is
$$(a, a\tan\theta)$$
and, the coordinates of Q is
$$(b, b\tan\theta)$$

Let R be (x, y). Let α and β be the given angles of directions for the straight lines PR and QR respectively.

Hence the equation of PR is
$$y - a\tan\theta = (x - a)\tan\alpha \tag{5.99}$$
Similarly the equation of QR is
$$y - b\tan\theta = (x - b)\tan\beta \tag{5.100}$$

(5.99) $\times\, b\, -$ (5.100) $\times\, a \implies$
$$(a - b)y = x(a\tan\beta - b\tan\alpha) + ab(\tan\alpha - \tan\beta).$$

This is the required locus, which is a straight line. ∎

Chapter 6

On Equations Representing Two Or More Straight Lines

6.1 Multiple Straight Lines and Included Angles

Find what straight lines are represented by the following equations and determine the angles between them.

§ **Problem 6.1.1.** $x^2 - 7xy + 12y^2 = 0$. ◊

§§ **Solution.** The given equation can be written as
$$(x - 3y)(x - 4y) = 0$$
This represents two straight lines whose equations are $x - 3y = 0$, and $x - 4y = 0$.
$$\therefore m_1 = \frac{1}{3} \text{ and } m_2 = \frac{1}{4}.$$
Hence, the angle between these two lines is
$$= \tan^{-1} \frac{\frac{1}{3} - \frac{1}{4}}{1 + \frac{1}{3} \cdot \frac{1}{4}} = \tan^{-1} \frac{1}{13}.$$

Alternatively, by *Art.* 110, the angle between these two lines can be computed as follows:
$$= \tan^{-1} \frac{2\sqrt{h^2 - ab}}{a + b}, \text{ where } h = -\frac{7}{2}, \text{ and } a = 1, b = 12$$
$$= \tan^{-1} \frac{2\sqrt{\left(\frac{7}{2}\right)^2 - 1 \cdot 12}}{1 + 12} = \tan^{-1} \frac{1}{13}, \text{ taking the +ve sign.}$$

■

6.1. Multiple Straight Lines and Included Angles

§ Problem 6.1.2. $4x^2 - 24xy + 11y^2 = 0$. ◊

§§ Solution. The given equation can be written as
$$(2x - y)(2x - 11y) = 0$$
This represents two straight lines whose equations are $2x - y = 0$, and $2x - 11y = 0$.
$$\therefore m_1 = 2 \text{ and } m_2 = \frac{2}{11}.$$
Hence, the angle between these two lines is
$$= \tan^{-1} \frac{2 - \frac{2}{11}}{1 + 2 \cdot \frac{2}{11}} = \tan^{-1} \frac{20}{15} = \tan^{-1} \frac{4}{3}.$$

Alternatively, by *Art.* 110, the angle between these two lines can be computed as follows:
$$= \tan^{-1} \frac{2\sqrt{h^2 - ab}}{a + b}, \text{ where } h = -12, \text{ and } a = 4, b = 11$$
$$= \tan^{-1} \frac{2\sqrt{(12)^2 - 4 \cdot 11}}{4 + 11} = \tan^{-1} \frac{20}{15} = \tan^{-1} \frac{4}{3}, \text{ taking the +ve sign.} \blacksquare$$

§ Problem 6.1.3. $33x^2 - 71xy - 14y^2 = 0$. ◊

§§ Solution. The given equation can be written as
$$(3x - 7y)(11x + 2y) = 0$$
This represents two straight lines whose equations are $3x - 7y = 0$, and $11x + 2y = 0$.
$$\therefore m_1 = \frac{3}{7} \text{ and } m_2 = -\frac{11}{2}.$$
Hence, the angle between these two lines is
$$= \tan^{-1} \frac{\frac{3}{7} - \left(-\frac{11}{2}\right)}{1 + \frac{3}{7} \cdot \left(-\frac{11}{2}\right)} = \tan^{-1} \frac{83}{19}.$$

Alternatively, by *Art.* 110, the angle between these two lines can be computed as follows:
$$= \tan^{-1} \frac{2\sqrt{h^2 - ab}}{a + b}, \text{ where } h = -\frac{71}{2}, \text{ and } a = 33, b = -14$$
$$= \tan^{-1} \frac{2\sqrt{\left(\frac{71}{2}\right)^2 - 33 \cdot (-14)}}{33 - 14} = \tan^{-1} \frac{83}{19}, \text{ taking the +ve sign.} \blacksquare$$

§ Problem 6.1.4. $x^3 - 6x^2 + 11x - 6 = 0$. ◊

§§ Solution. The given equation can be written as
$$(x - 1)(x - 2)(x - 3) = 0$$
This represents three straight lines whose equations are $x - 1 = 0$, $x - 2 = 0$, and $x - 3 = 0$.

It is easy to see that these three lines are parallel to the y-axis, hence the angle between these lines is zero. $\blacksquare$

§ Problem 6.1.5. $y^2 - 16 = 0$. ◊

6.1. Multiple Straight Lines and Included Angles

§§ Solution. The given equation can be written as
$$(y-4)(y+4) = 0$$
This represents two straight lines whose equations are $y - 4 = 0$, and $y + 4 = 0$.

It is easy to see that these two lines are parallel to the x-axis, hence the angle between these lines is zero. ∎

§ Problem 6.1.6. $y^3 - xy^2 - 14x^2y + 24x^3 = 0$. ◊

§§ Solution. The given equation can be written as
$$(y + 4x)(y - 3x)(y - 2x) = 0$$
This represents three straight lines whose equations are $y + 4x = 0$, $y - 3x = 0$, and $y - 2x = 0$.

$$\therefore m_1 = -4,\ m_2 = 3 \text{ and } m_3 = 2.$$

Hence, the angle between third and first lines is
$$= \tan^{-1} \frac{2+4}{1 - 2 \cdot 4} = \tan^{-1} \frac{-6}{7}.$$

Hence, the angle between second and third lines is
$$= \tan^{-1} \frac{3-2}{1 + 3 \cdot 2} = \tan^{-1} \frac{1}{7}. \quad \blacksquare$$

§ Problem 6.1.7. $x^2 + 2xy \sec\theta + y^2 = 0$. ◊

§§ Solution. The given equation can be written as
$$\left(\frac{y}{x}\right)^2 + 2\sec\theta\left(\frac{y}{x}\right) + 1 = 0$$
$$\therefore \frac{y}{x} = \frac{-2\sec\theta \pm \sqrt{4\sec^2\theta - 4}}{2} = -\sec\theta \pm \sqrt{\sec^2\theta - 1} = -\sec\theta \pm \tan\theta$$
$$\therefore \frac{y}{x} = \frac{-1 \pm \sin\theta}{\cos\theta}$$

Hence the two lines are
$$x(1 + \sin\theta) + y\cos\theta = 0, \text{ and}$$
$$x(1 - \sin\theta) + y\cos\theta = 0$$

By *Art.* 110, the angle between these two lines can be computed as follows:
$$= \tan^{-1} \frac{2\sqrt{h^2 - ab}}{a + b}, \text{ where } h = \sec\theta, \text{ and } a = 1,\ b = 1$$
$$= \tan^{-1} \frac{2\sqrt{\sec^2\theta - 1}}{1 + 1} = \tan^{-1} \tan\theta = \theta. \quad \blacksquare$$

§ Problem 6.1.8. $x^2 + 2xy \cot\theta + y^2 = 0$. ◊

§§ Solution. The given equation can be written as
$$\left(\frac{y}{x}\right)^2 + 2\cot\theta\left(\frac{y}{x}\right) + 1 = 0$$
$$\therefore \frac{y}{x} = \frac{-2\cot\theta \pm \sqrt{4\cot^2\theta - 4}}{2} = -\cot\theta \pm \sqrt{\cot^2\theta - 1}$$
$$= -\frac{\cos\theta}{\sin\theta} \pm \sqrt{\left(\frac{\cos^2\theta - \sin^2\theta}{\sin^2\theta}\right)}$$
$$\therefore \frac{y}{x} = \frac{-\cos\theta \pm \sqrt{\cos 2\theta}}{\sin\theta}$$

6.1. Multiple Straight Lines and Included Angles

Hence the two lines are
$$y\sin\theta + x(\cos\theta + \sqrt{\cos 2\theta}) = 0, \text{ and}$$
$$y\sin\theta + x(\cos\theta - \sqrt{\cos 2\theta}) = 0.$$

By *Art.* 110, the angle between these two lines can be computed as follows:
$$= \tan^{-1}\frac{2\sqrt{h^2 - ab}}{a + b}, \text{ where } h = \cot\theta, \text{ and } a = 1, b = 1$$
$$= \tan^{-1}\frac{2\sqrt{\cot^2\theta - 1}}{1 + 1} = \tan^{-1}\left(\operatorname{cosec}\theta\sqrt{\cos 2\theta}\right). \blacksquare$$

§ **Problem 6.1.9.** *Find the equations of the straight lines bisecting the angles between the pairs of straight lines given in 6.1.2, 6.1.3, 6.1.7 and 6.1.8.* ◊

§§ **Solution.** By *Art.* 112, the required equations are given by
$$\frac{x^2 - y^2}{a - b} = \frac{xy}{h}.$$

Hence the required equations for $4x^2 - 24xy + 11y^2 = 0$ are given by
$$\frac{x^2 - y^2}{4 - 11} = \frac{xy}{-12}$$
$$\therefore 12x^2 - 7xy - 12y^2 = 0.$$

The required equations for $33x^2 - 71xy - 14y^2 = 0$ are given by
$$\frac{x^2 - y^2}{33 + 14} = \frac{xy}{-\frac{71}{2}}$$
$$\therefore 71x^2 + 94xy - 71y^2 = 0.$$

The required equations for $x^2 + 2xy\sec\theta + y^2 = 0$ are given by
$$\frac{x^2 - y^2}{1 - 1} = \frac{xy}{\sec\theta}$$
$$\therefore x^2 - y^2 = 0.$$

The required equations for $x^2 + 2xy\cot\theta + y^2 = 0$ are given by
$$\frac{x^2 - y^2}{1 - 1} = \frac{xy}{\cot\theta}$$
$$\therefore x^2 - y^2 = 0. \blacksquare$$

§ **Problem 6.1.10.** *Show that the two straight lines*
$$x^2\left(\tan^2\theta + \cos^2\theta\right) - 2xy\tan\theta + y^2\sin^2\theta = 0$$
make with the axis of x angles such that the difference of their tangents is 2. ◊

§§ **Solution.** By *Art.* 112 :
$$\tan\theta_1 + \tan\theta_2 = -\frac{2h}{b}, \text{ where } h = -\tan\theta, \text{ and } b = \sin^2\theta,$$
$$\therefore \tan\theta_1 + \tan\theta_2 = \frac{2\tan\theta}{\sin^2\theta} = 2\sec\theta\operatorname{cosec}\theta.$$
and $\tan\theta_1 \cdot \tan\theta_2 = \frac{a}{b}$, where $a = \tan^2\theta + \cos^2\theta$
$$\therefore \tan\theta_1 \cdot \tan\theta_2 = \frac{\tan^2\theta + \cos^2\theta}{\sin^2\theta} = \sec^2\theta + \cot^2\theta.$$
$$\therefore \tan\theta_1 - \tan\theta_2 = \sqrt{(\tan\theta_1 + \tan\theta_2)^2 - 4\tan\theta_1 \cdot \tan\theta_2}$$

6.1. Multiple Straight Lines and Included Angles

$$= \sqrt{4\sec^2\theta\csc^2\theta - 4(\sec^2\theta + \cot^2\theta)}$$
$$= 2\sqrt{\sec^2\theta(\csc^2\theta - 1) - \cot^2\theta}$$
$$= 2\sqrt{\sec^2\theta \cdot \cot^2\theta - \cot^2\theta} = 2\sqrt{(\sec^2\theta - 1)\cot^2\theta}$$
$$= 2\sqrt{\tan^2\theta \cdot \cot^2\theta} = 2.$$

Alternative Solution :

$$\tan\theta_1 + \tan\theta_2 = -\frac{2h}{b}, \text{ where } h = -\tan\theta, \text{ and } b = \sin^2\theta,$$
$$\therefore \tan\theta_1 + \tan\theta_2 = \frac{2\tan\theta}{\sin^2\theta}$$
$$= \frac{2}{\sin\theta \cdot \cos\theta} = \frac{2(\sin^2\theta + \cos^2\theta)}{\sin\theta \cdot \cos\theta} = 2(\tan\theta + \cot\theta).$$
$$\tan\theta_1 \cdot \tan\theta_2 = \frac{a}{b}, \text{ where } a = \tan^2\theta + \cos^2\theta$$
$$\therefore \tan\theta_1 \cdot \tan\theta_2 = \frac{\tan^2\theta + \cos^2\theta}{\sin^2\theta} = \sec^2\theta + \cot^2\theta = 1 + \tan^2\theta + \cot^2\theta.$$
$$\therefore \tan\theta_1 - \tan\theta_2 = \sqrt{(\tan\theta_1 + \tan\theta_2)^2 - 4\tan\theta_1 \cdot \tan\theta_2}$$
$$= 2\sqrt{(\tan\theta + \cot\theta)^2 - (1 + \tan^2\theta + \cot^2\theta)} = 2.$$

Alternative Solution :

The given equation is
$$x^2(\tan^2\theta + \cos^2\theta) - 2xy\tan\theta + y^2\sin^2\theta = 0$$
$$\therefore y^2 - 2xy\frac{\tan\theta}{\sin^2\theta} + x^2\frac{(\tan^2\theta + \cos^2\theta)}{\sin^2\theta} = 0 \tag{6.1}$$
Suppose this represents the following two lines :
$$y - \tan\theta_1 x = 0, \text{ and } y - \tan\theta_2 x = 0$$
$$\therefore (y - \tan\theta_1 x)(y - \tan\theta_2 x) = 0$$
$$\therefore y^2 - (\tan\theta_1 + \tan\theta_2)x + (\tan\theta_1 \cdot \tan\theta_2)x^2 = 0. \tag{6.2}$$
Comparing (6.1) and (6.2) :
$$\tan\theta_1 + \tan\theta_2 = \frac{2\tan\theta}{\sin^2\theta}, \text{ and}$$
$$\tan\theta_1 \cdot \tan\theta_2 = \frac{(\tan^2\theta + \cos^2\theta)}{\sin^2\theta}.$$
Rest follows as described in the previous solutions. ∎

§ Problem 6.1.11. *Prove that the two straight lines*
$$(x^2 + y^2)\left(\cos^2\theta\sin^2\alpha + \sin^2\theta\right) = (x\tan\alpha - y\sin\theta)^2$$
include an angle 2α. ◇

§§ Solution. The given equation is
$$(x^2 + y^2)(\cos^2\theta\sin^2\alpha + \sin^2\theta) = (x\tan\alpha - y\sin\theta)^2$$
$$\therefore (x^2 + y^2)\left\{\cos^2\theta\sin^2\alpha + \sin^2\theta(\sin^2\alpha + \cos^2\alpha)\right\} = (x\tan\alpha - y\sin\theta)^2$$
$$\therefore (x^2 + y^2)\left\{(\cos^2\theta + \sin^2\theta)\sin^2\alpha + \sin^2\theta\cos^2\alpha\right\} = (x\tan\alpha - y\sin\theta)^2$$
$$\therefore (x^2 + y^2)\left\{\sin^2\alpha + \sin^2\theta\cos^2\alpha\right\} = (x\tan\alpha - y\sin\theta)^2$$
Dividing both sides by $\cos^2\alpha$, we get
$$(x^2 + y^2)(\tan^2\alpha + \sin^2\theta) = \sec^2\alpha(x\tan\alpha - y\sin\theta)^2$$
$$\therefore (x^2 + y^2)(\tan^2\alpha + \sin^2\theta) = (1 + \tan^2\alpha)(x\tan\alpha - y\sin\theta)^2$$

6.1. Multiple Straight Lines and Included Angles 150

$\therefore x^2 \sin^2 \theta + y^2 \tan^2 \alpha + 2xy \tan\alpha \sin\theta = \tan^2\alpha(x\tan\alpha - y\sin\theta)^2$

$\therefore (x\sin\theta + y\tan\alpha)^2 = \{\tan\alpha(x\tan\alpha - y\sin\theta)\}^2$

$\therefore x\sin\theta + y\tan\alpha = \pm\tan\alpha(x\tan\alpha - y\sin\theta)$

$\therefore y\tan\alpha(1 \pm \sin\theta) = x(-\sin\theta \pm \tan^2\alpha)$

$$\therefore y = \frac{-\sin\theta \pm \tan^2\alpha}{\tan\alpha(1 \pm \sin\theta)} x$$

This represents two straight lines with slopes as

$$m_1 = \frac{-\sin\theta + \tan^2\alpha}{\tan\alpha(1+\sin\theta)}, \text{ and } m_2 = \frac{-\sin\theta - \tan^2\alpha}{\tan\alpha(1-\sin\theta)}$$

Hence the angle between these lines is computed as follows

$$= \tan^{-1}\frac{m_1 - m_2}{1 + m_1 m_2} = \tan^{-1}\frac{\dfrac{-\sin\theta + \tan^2\alpha}{\tan\alpha(1+\sin\theta)} + \dfrac{\sin\theta + \tan^2\alpha}{\tan\alpha(1-\sin\theta)}}{1 - \dfrac{\tan^4\alpha - \sin^2\theta}{\tan^2\alpha(1-\sin^2\theta)}}$$

$$= \tan^{-1}\frac{2\tan\alpha(\tan^2\alpha + \sin^2\theta)}{(1-\tan^2\alpha)(\tan^2\alpha + \sin^2\theta)}$$

$$= \tan^{-1}\frac{2\tan\alpha}{1-\tan^2\alpha} = \tan^{-1}\tan 2\alpha = 2\alpha.$$

Alternative Solution :

The given equation is

$(x^2 + y^2)(\cos^2\theta \sin^2\alpha + \sin^2\theta) = (x\tan\alpha - y\sin\theta)^2$

$\therefore x^2(\cos^2\theta \sin^2\alpha + \sin^2\theta - \tan^2\alpha)$
$+ y^2(\cos^2\theta \sin^2\alpha + \sin^2\theta - \sin^2\theta) + 2xy\tan\alpha\sin\theta = 0$

$\therefore x^2(\cos^2\theta \sin^2\alpha + \sin^2\theta - \tan^2\alpha) + y^2\cos^2\theta\sin^2\alpha +$
$2xy\tan\alpha\sin\theta = 0$

Comparing this with $ax^2 + 2hxy + by^2 = 0$:

$a = \cos^2\theta\sin^2\alpha + \sin^2\theta - \tan^2\alpha,$
$b = \cos^2\theta\sin^2\alpha, \ h = 2\tan\alpha\sin\theta$

By *Art.* 110, the angle between the lines $= \tan^{-1}\dfrac{2\sqrt{h^2 - ab}}{a+b}$.

Le us compute $h^2 - ab$:

$h^2 - ab$
$= \tan^2\alpha\sin^2\theta - (\cos^2\theta\sin^2\alpha + \sin^2\theta - \tan^2\alpha)\cos^2\theta\sin^2\alpha$
$= \tan^2\alpha$
$(\sin^2\theta - \cos^4\theta\sin^2\alpha\cos^2\alpha - \sin^2\theta\cos^2\theta\cos^2\alpha + \cos^2\theta\sin^2\alpha)$
$= \tan^2\alpha(\sin^2\theta + \cos^2\theta\sin^2\alpha)(1 - \cos^2\theta\cos^2\alpha)$
$= \tan^2\alpha(\sin^2\theta + \cos^2\theta\sin^2\alpha)(\sin^2\theta + \cos^2\theta - \cos^2\theta\cos^2\alpha)$
$= \tan^2\alpha(\sin^2\theta + \cos^2\theta\sin^2\alpha)\left\{\sin^2\theta + \cos^2\theta(1 - \cos^2\alpha)\right\}$
$= \tan^2\alpha(\sin^2\theta + \cos^2\theta\sin^2\alpha)(\sin^2\theta + \cos^2\theta\sin^2\alpha)$
$= \left\{\tan\alpha(\sin^2\theta + \cos^2\theta\sin^2\alpha)\right\}^2$

Le us compute $a + b$:
$a + b = 2\cos^2\theta\sin^2\alpha + \sin^2\theta - \tan^2\alpha$
$= \sec^2\alpha(\cos^2\theta\sin^2\alpha\cos^2\alpha + \sin^2\theta\cos^2\alpha - \sin^2\alpha)$

6.1. Multiple Straight Lines and Included Angles

$$= \sec^2\alpha(\cos^2\theta\sin^2\alpha\cos^2\alpha + \sin^2\theta\cos^2\alpha - (\cos^2\theta + \sin^2\theta)\sin^2\alpha)$$
$$= \sec^2\alpha\left\{\sin^2\theta(\cos^2\alpha - \sin^2\alpha) + \cos^2\theta\sin^2\alpha(2\cos^2\alpha - 1)\right\}$$
$$= \sec^2\alpha\left\{\sin^2\theta(\cos^2\alpha - \sin^2\alpha) + \cos^2\theta\sin^2\alpha(\cos^2\alpha - \sin^2\alpha)\right\}$$
$$= \sec^2\alpha(\cos^2\alpha - \sin^2\alpha)(\sin^2\theta + \cos^2\theta\sin^2\alpha)$$
$$= (1 - \tan^2\alpha)(\sin^2\theta + \cos^2\theta\sin^2\alpha)$$

Hence the angle between the lines
$$= \tan^{-1}\frac{2\sqrt{h^2 - ab}}{a + b}$$
$$= \tan^{-1}\frac{2\tan\alpha(\sin^2\theta + \cos^2\theta\sin^2\alpha)}{(1 - \tan^2\alpha)(\sin^2\theta + \cos^2\theta\sin^2\alpha)}$$
$$= \tan^{-1}\frac{2\tan\alpha}{1 - \tan^2\alpha}$$
$$= \tan^{-1}\tan 2\alpha = 2\alpha. \qquad \blacksquare$$

§ Problem 6.1.12. *Prove that the two straight lines*
$$x^2\sin^2\alpha\cos^2\theta + 4xy\sin\alpha\sin\theta + y^2\left[4\cos\alpha - (1+\cos\alpha)^2\cos^2\theta\right] = 0$$
include an angle α. $\diamond$

§§ Solution. The given equation is
$$x^2\sin^2\alpha\cos^2\theta + 4xy\sin\alpha\sin\theta + y^2[4\cos\alpha - (1+\cos\alpha)^2\cos^2\theta] = 0$$
$$\therefore x^2\sin^2\alpha(1 - \sin^2\theta) + 4xy\sin\alpha\sin\theta +$$
$$y^2\left[(1+\cos\alpha)^2 - (1-\cos\alpha)^2 - (1+\cos\alpha)^2\cos^2\theta\right] = 0$$
$$\therefore x^2\sin^2\alpha - x^2\sin^2\alpha\sin^2\theta + 2xy\sin\alpha\sin\theta[(1+\cos\alpha) + (1-\cos\alpha)] +$$
$$y^2\left[(1+\cos\alpha)^2(1-\cos^2\theta) - (1-\cos\alpha)^2\right] = 0$$
$$\therefore [x\sin\alpha]^2 + 2(x\sin\alpha)y(1+\cos\alpha)\sin\theta + [y(1+\cos\alpha)\sin\theta]^2 =$$
$$[x\sin\alpha\sin\theta]^2 - 2(x\sin\alpha\sin\theta)y(1-\cos\alpha) + [y(1-\cos\alpha)]^2$$
$$\therefore [x\sin\alpha + y(1+\cos\alpha)\sin\theta]^2 = [x\sin\alpha\sin\theta - y(1-\cos\alpha)]^2$$
$$\therefore x\sin\alpha + y(1+\cos\alpha)\sin\theta = \pm[x\sin\alpha\sin\theta - y(1-\cos\alpha)]$$

Dividing both sides by $(1+\cos\alpha)$:
$$x\frac{\sin\alpha}{1+\cos\alpha} + y\sin\theta = \pm\left[x\sin\theta\frac{\sin\alpha}{1+\cos\alpha} - y\frac{1-\cos\alpha}{1+\cos\alpha}\right]$$
$$\because \tan\frac{\alpha}{2} = \frac{\sin\alpha}{1+\cos\alpha} = \pm\sqrt{\frac{1-\cos\alpha}{1+\cos\alpha}}$$
$$\therefore x\tan\frac{\alpha}{2} + y\sin\theta = \pm\left[x\sin\theta\tan\frac{\alpha}{2} - y\tan^2\frac{\alpha}{2}\right]$$
$$\therefore y\left(\sin\theta \pm \tan^2\frac{\alpha}{2}\right) = x\tan\frac{\alpha}{2}(\pm\sin\theta - 1)$$
$$\therefore y = \frac{\tan\dfrac{\alpha}{2}(\pm\sin\theta - 1)}{\left(\sin\theta \pm \tan^2\dfrac{\alpha}{2}\right)}x$$

This represents two straight lines with slopes as
$$m_1 = \frac{\tan\dfrac{\alpha}{2}(\sin\theta - 1)}{\left(\sin\theta + \tan^2\dfrac{\alpha}{2}\right)}, \quad \text{and} \quad m_2 = \frac{\tan\dfrac{\alpha}{2}(-\sin\theta - 1)}{\left(\sin\theta - \tan^2\dfrac{\alpha}{2}\right)}$$

6.2. General Equation of The Second Degree

Hence the angle between these lines is computed as follows

$$= \tan^{-1} \frac{m_1 - m_2}{1 + m_1 m_2} = \tan^{-1} \frac{\dfrac{\tan \dfrac{\alpha}{2}(\sin\theta - 1)}{\left(\sin\theta + \tan^2 \dfrac{\alpha}{2}\right)} + \dfrac{\tan \dfrac{\alpha}{2}(\sin\theta + 1)}{\left(\sin\theta - \tan^2 \dfrac{\alpha}{2}\right)}}{1 + \dfrac{\tan^2 \dfrac{\alpha}{2}(\sin^2\theta - 1)}{\tan^4 \dfrac{\alpha}{2} - \sin^2\theta}}$$

$$= \tan^{-1} \frac{2\tan \dfrac{\alpha}{2}\left(\tan^2 \dfrac{\alpha}{2} + \sin^2\theta\right)}{\left(1 - \tan^2 \dfrac{\alpha}{2}\right)\left(\tan^2 \dfrac{\alpha}{2} + \sin^2\theta\right)}$$

$$= \tan^{-1} \frac{2\tan \dfrac{\alpha}{2}}{\left(1 - \tan^2 \dfrac{\alpha}{2}\right)} = \tan^{-1}\tan\alpha = \alpha.$$

Alternative Solution :

Comparing the given equation with $ax^2 + 2hxy + by^2 = 0$:
$a = \sin^2\alpha\cos^2\theta$, $b = 4\cos\alpha - (1 + \cos\alpha)^2\cos^2\theta$, $h = 2\sin\alpha\sin\theta$.

By $Art.\,110$, the angle between the lines $= \tan^{-1}\dfrac{2\sqrt{h^2 - ab}}{a + b}$.

Le us compute $h^2 - ab$:

$$\begin{aligned}
h^2 - ab &= 4\sin^2\alpha\sin^2\theta - \sin^2\alpha\cos^2\theta\left\{4\cos\alpha - (1 + \cos\alpha)^2\cos^2\theta\right\} \\
&= \sin^2\alpha\left\{4\sin^2\theta - 4\cos\alpha\cos^2\theta + (1 + \cos\alpha)^2\cos^4\theta\right\} \\
&= \sin^2\alpha\left\{4 - 4\cos^2\theta - 4\cos\alpha\cos^2\theta + (1 + \cos\alpha)^2\cos^4\theta\right\} \\
&= \sin^2\alpha\left\{4 - 4\cos^2\theta(1 + \cos\alpha) + (1 + \cos\alpha)^2\cos^4\theta\right\} \\
&= \sin^2\alpha\left\{2 - (1 + \cos\alpha)\cos^2\theta\right\}^2
\end{aligned}$$

Le us compute $a + b$:
$$\begin{aligned}
a + b &= \sin^2\alpha\cos^2\theta + 4\cos\alpha - (1 + \cos\alpha)^2\cos^2\theta \\
&= 4\cos\alpha - (1 + \cos^2\alpha + 2\cos\alpha)\cos^2\theta + \sin^2\alpha\cos^2\theta \\
&= 4\cos\alpha - \cos^2\theta(1 + \cos^2\alpha + 2\cos\alpha - \sin^2\alpha) \\
&= 4\cos\alpha - \cos^2\theta(2\cos^2\alpha + 2\cos\alpha) \\
&= 2\cos\alpha\left[2 - (1 + \cos\alpha)\cos^2\theta\right]
\end{aligned}$$

Hence the angle between the lines

$$= \tan^{-1}\frac{2\sqrt{h^2 - ab}}{a + b}$$
$$= \tan^{-1}\frac{2\sin\alpha\left\{2 - (1 + \cos\alpha)\cos^2\theta\right\}}{2\cos\alpha\left[2 - (1 + \cos\alpha)\cos^2\theta\right]} = \tan^{-1}\tan\alpha = \alpha. \quad\blacksquare$$

6.2 General Equation of The Second Degree

Prove that the following equations represent two straight lines; find also their point of intersection and the angle between them.

6.2. General Equation of The Second Degree

§ Problem 6.2.1. $6y^2 - xy - x^2 + 30y + 36 = 0$. ◊

§§ Solution. The given equation is
$$-x^2 + 6y^2 - xy + 30y + 36 = 0. \tag{6.3}$$

By *Art.* 118, the Discriminant
$$\Delta = \begin{vmatrix} a & h & g \\ h & b & f \\ g & f & c \end{vmatrix}$$

where $a = -1$, $b = 6$, $c = 36$, $h = -\dfrac{1}{2}$, $g = 0$, $f = 15$.

$$\therefore \Delta = \begin{vmatrix} -1 & -\frac{1}{2} & 0 \\ -\frac{1}{2} & 6 & 15 \\ 0 & 15 & 36 \end{vmatrix}$$

$$= -1\left(6 \cdot 36 - 15^2\right) - \left(-\frac{1}{2}\right) \cdot \left(-\frac{1}{2} \cdot 36\right) = 9 - 9 = 0.$$

The equation therefore represents two straight lines.

The terms of the second degree in the equation (6.3) break up into the factors as
$$6y^2 - xy - x^2 = (3y + x)(2y - x).$$

If $(3y + x + A)(2y - x + B) \equiv 6y^2 - xy - x^2 + 30y + 36$, then
$$B - A = 0, \ 2A + 3B = 30, \ AB = 36.$$
$$\therefore A = B = 6.$$

Hence the two equations are as follows
$$3y + x + 6 = 0, \text{ and}$$
$$2y - x + 6 = 0.$$

Solving these equations, $x = \dfrac{6}{5}$, $y = -\dfrac{12}{5}$, hence the point of intersection is $\left(\dfrac{6}{5}, -\dfrac{12}{5}\right)$.

And the angle is $= \tan^{-1} \dfrac{\frac{1}{2} + \frac{1}{3}}{1 - \frac{1}{2} \cdot \frac{1}{3}} = \tan^{-1} 1 = 45°$.

Alternatively, the angle is $= \tan^{-1} \dfrac{2\sqrt{h^2 - ab}}{a + b} = \tan^{-1} \dfrac{2\sqrt{\frac{1}{4} + 6}}{-1 + 6} = \tan^{-1} 1 = 45°$. ∎

§ Problem 6.2.2. $x^2 - 5xy + 4y^2 + x + 2y - 2 = 0$. ◊

§§ Solution. The given equation is
$$x^2 - 5xy + 4y^2 + x + 2y - 2 = 0. \tag{6.4}$$

By *Art.* 118, the Discriminant
$$\Delta = \begin{vmatrix} a & h & g \\ h & b & f \\ g & f & c \end{vmatrix}$$

where $a = 1$, $b = 4$, $c = -2$, $h = -\dfrac{5}{2}$, $g = \dfrac{1}{2}$, $f = 1$.

$$\therefore \Delta = \begin{vmatrix} 1 & -\frac{5}{2} & \frac{1}{2} \\ -\frac{5}{2} & 4 & 1 \\ \frac{1}{2} & 1 & -2 \end{vmatrix} = -9 + \frac{5}{2} \cdot \left(5 - \frac{1}{2}\right) + \frac{1}{2} \cdot \left(\frac{-5}{2} - 2\right) = 0.$$

The equation therefore represents two straight lines.

6.2. General Equation of The Second Degree

The terms of the second degree in the equation (6.4) break up into the factors as
$$x^2 - 5xy + 4y^2 = (x - 4y)(x - y).$$
If $(x - 4y + A)(x - y + B) \equiv x^2 - 5xy + 4y^2 + x + 2y - 2$, then
$$A + B = 1, \ A + 4B = -2, \ AB = -2.$$
$$\therefore A = 2, \ B = -1.$$
Hence the two equations are as follows
$$x - 4y + 2 = 0, \text{ and}$$
$$x - y - 1 = 0.$$
Solving these equations, $x = 2$, $y = 1$, hence the point of intersection is $(2, 1)$.

And the angle is $= \tan^{-1} \dfrac{1 - \frac{1}{4}}{1 + 1 \cdot \frac{1}{4}} = \tan^{-1} \dfrac{3}{5}$.

Alternatively, the angle is $= \tan^{-1} \dfrac{2\sqrt{h^2 - ab}}{a+b} = \tan^{-1} \dfrac{2\sqrt{\frac{25}{4} - 4}}{1 + 4} = \tan^{-1} \dfrac{3}{5}$. ∎

§ Problem 6.2.3. $3y^2 - 8xy - 3x^2 - 29x + 3y - 18 = 0$. ◊

§§ Solution. The given equation is
$$3y^2 - 8xy - 3x^2 - 29x + 3y - 18 = 0. \quad (6.5)$$
By *Art.* 118, the Discriminant
$$\Delta = \begin{vmatrix} a & h & g \\ h & b & f \\ g & f & c \end{vmatrix}$$
where $a = -3$, $b = 3$, $c = -18$, $h = -4$, $g = -\dfrac{29}{2}$, $f = \dfrac{3}{2}$.
$$\therefore \Delta = \begin{vmatrix} -3 & -4 & -\frac{29}{2} \\ -4 & 3 & \frac{3}{2} \\ -\frac{29}{2} & \frac{3}{2} & -18 \end{vmatrix}$$
$$= -3\left(-54 - \dfrac{9}{4}\right) + 4\left(72 + \dfrac{87}{4}\right) - \dfrac{29}{2}\left(-6 + \dfrac{87}{2}\right) = 0.$$
The equation therefore represents two straight lines.

The terms of the second degree in the equation (6.5) break up into the factors as
$$3y^2 - 8xy - 3x^2 = (3y + x)(y - 3x).$$
If $(3y + x + A)(y - 3x + B) \equiv 3y^2 - 8xy - 3x^2 - 29x + 3y - 18$, then
$$3A - B = 29, \ 3B + A = 3, \ AB = -18.$$
$$\therefore A = 9, \ B = -2.$$
Hence the two equations are as follows
$$3y + x + 9 = 0, \text{ and}$$
$$y - 3x - 2 = 0.$$
Solving these equations, $x = -\dfrac{3}{2}$, $y = -\dfrac{5}{2}$, hence the point of intersection is $\left(-\dfrac{3}{2}, -\dfrac{5}{2}\right)$.

And the angle is $= \tan^{-1} \dfrac{3 + \frac{1}{3}}{1 - 3 \cdot \frac{1}{3}} = \tan^{-1} \infty = 90°$.

6.2. General Equation of The Second Degree

Alternatively, the angle is $= \tan^{-1} \dfrac{2\sqrt{h^2 - ab}}{a + b} = \tan^{-1} \dfrac{2\sqrt{16 + 9}}{-3 + 3} = \tan^{-1} \infty = 90°$. ∎

§ Problem 6.2.4. $3y^2 - 8xy - 3x^2 - 29x + 3y - 18 = 0$. ◊

§§ Solution. The given equation is
$$y^2 + xy - 2x^2 - 5x - y - 2 = 0. \qquad (6.6)$$
By *Art.* 118, the Discriminant
$$\Delta = \begin{vmatrix} a & h & g \\ h & b & f \\ g & f & c \end{vmatrix}$$
where $a = -2$, $b = 1$, $c = -2$, $h = \dfrac{1}{2}$, $g = -\dfrac{5}{2}$, $f = -\dfrac{1}{2}$.

$$\therefore \Delta = \begin{vmatrix} -2 & \tfrac{1}{2} & -\tfrac{5}{2} \\ \tfrac{1}{2} & 1 & -\tfrac{1}{2} \\ -\tfrac{5}{2} & -\tfrac{1}{2} & -2 \end{vmatrix}$$
$$= -2\left(-2 - \dfrac{1}{4}\right) - \dfrac{1}{2}\left(-1 - \dfrac{5}{4}\right) - \dfrac{5}{2}\left(-\dfrac{1}{4} + \dfrac{5}{2}\right) = 0.$$

The equation therefore represents two straight lines.

The terms of the second degree in the equation (6.6) break up into the factors as
$$y^2 + xy - 2x^2 = (y + 2x)(y - x).$$
If $(y + 2x + A)(y - x + B) \equiv y^2 + xy - 2x^2 - 5x - y - 2$, then
$$2B - A = -5, \ A + B = -1, \ AB = -2.$$
$$\therefore A = 1, \ B = -2.$$
Hence the two equations are as follows
$$y + 2x + 1 = 0, \text{ and}$$
$$y - x - 2 = 0.$$
Solving these equations, $x = -1$, $y = 1$, hence the point of intersection is $(-1, 1)$.

And the angle is $= \tan^{-1} \dfrac{-2 - 1}{1 - 2} = \tan^{-1} 3$.

Alternatively, the angle is $= \tan^{-1} \dfrac{2\sqrt{h^2 - ab}}{a + b} = \tan^{-1} \dfrac{2\sqrt{\tfrac{1}{4} + 2}}{-2 + 1} = \tan^{-1} 3$. (Ignoring the -ve sign). ∎

§ Problem 6.2.5. *Prove that the equation*
$$x^2 + 6xy + 9y^2 + 4x + 12y - 5 = 0$$
represents two parallel lines. ◊

§§ Solution. The given equation is
$$x^2 + 6xy + 9y^2 + 4x + 12y - 5 = 0. \qquad (6.7)$$
By *Art.* 118, the Discriminant
$$\Delta = \begin{vmatrix} a & h & g \\ h & b & f \\ g & f & c \end{vmatrix}$$
where $a = 1$, $b = 9$, $c = -5$, $h = 3$, $g = 2$, $f = 6$.
$$\therefore \Delta = \begin{vmatrix} 1 & 3 & 2 \\ 3 & 9 & 6 \\ 2 & 6 & -5 \end{vmatrix} = 0.$$

6.2. General Equation of The Second Degree

The equation therefore represents two straight lines.

The terms of the second degree in the equation (6.7) break up into the factors as
$$x^2 + 6xy + 9y^2 = (x+3y)^2.$$
Hence the lines are parallel.

Alternatively, the angle is $= \tan^{-1} \dfrac{2\sqrt{h^2-ab}}{a+b} = \tan^{-1} \dfrac{2\sqrt{9-9}}{1+9} = 0.$
Hence the lines are parallel. ∎

Find the value of k ; so that the following equations may represent pairs of straight lines :

§ Problem 6.2.6. $6x^2 + 11xy - 10y^2 + x + 31y + k = 0.$ ◇

§§ Solution. The given equation is
$$6x^2 + 11xy - 10y^2 + x + 31y + k = 0.$$
By *Art.* 118, the Discriminant of this equation should be zero, i.e.,
$$\Delta = \begin{vmatrix} a & h & g \\ h & b & f \\ g & f & c \end{vmatrix} = 0.$$
where $a=6$, $b=-10$, $c=k$, $h=\dfrac{11}{2}$, $g=\dfrac{1}{2}$, $f=\dfrac{31}{2}$.
$$\therefore \Delta = \begin{vmatrix} 6 & \frac{11}{2} & \frac{1}{2} \\ \frac{11}{2} & -10 & \frac{31}{2} \\ \frac{1}{2} & \frac{31}{2} & k \end{vmatrix} = 0.$$
$$\therefore 6\left\{-10k - \left(\frac{31}{2}\right)^2\right\} - \frac{11}{2}\left(\frac{11k}{2} - \frac{31}{4}\right) + \frac{1}{2}\left(\frac{343}{4} + 5\right) = 0$$
$$\therefore -60k - \frac{2883}{4} - \frac{121k}{2} + \frac{341}{8} + \frac{341}{8} + \frac{5}{2} = 0$$
$$\therefore \frac{361k}{4} = -\frac{5415}{4}$$
$$\therefore k = -15. \quad ∎$$

§ Problem 6.2.7. $12x^2 - 10xy + 2y^2 + 11x - 5y + k = 0.$ ◇

§§ Solution. The given equation is
$$12x^2 - 10xy + 2y^2 + 11x - 5y + k = 0.$$
By *Art.* 118, the Discriminant of this equation should be zero, i.e.,
$$\Delta = \begin{vmatrix} a & h & g \\ h & b & f \\ g & f & c \end{vmatrix} = 0.$$
where $a=12$, $b=2$, $c=k$, $h=-5$, $g=\dfrac{11}{2}$, $f=-\dfrac{5}{2}$.
$$\therefore \Delta = \begin{vmatrix} 12 & -5 & \frac{11}{2} \\ -5 & 2 & -\frac{5}{2} \\ \frac{11}{2} & -\frac{5}{2} & k \end{vmatrix} = 0.$$
$$\therefore 12\left(2k - \frac{25}{4}\right) + 5\left(-5k + \frac{55}{4}\right) + \frac{11}{2}\left(\frac{25}{2} - 11\right) = 0$$
$$\therefore 24k - 75 - 25k + \frac{275}{2} - \frac{121}{2} = 0$$
$$\therefore -k + 2 = 0$$
$$\therefore k = 2. \quad ∎$$

6.2. General Equation of The Second Degree   157

§ **Problem 6.2.8.** $12x^2 + kxy + 2y^2 + 11x - 5y + 2 = 0$. ◊

§§ **Solution.** The given equation is
$$12x^2 + kxy + 2y^2 + 11x - 5y + 2 = 0.$$
By *Art.* 118, the Discriminant of this equation should be zero, i.e.,
$$\Delta = \begin{vmatrix} a & h & g \\ h & b & f \\ g & f & c \end{vmatrix} = 0.$$
where $a = 12$, $b = 2$, $c = 2$, $h = \dfrac{k}{2}$, $g = \dfrac{11}{2}$, $f = -\dfrac{5}{2}$.

$$\therefore \Delta = \begin{vmatrix} 12 & \frac{k}{2} & \frac{11}{2} \\ \frac{k}{2} & 2 & -\frac{5}{2} \\ \frac{11}{2} & -\frac{5}{2} & 2 \end{vmatrix} = 0.$$

$$\therefore 12\left(4 - \frac{25}{4}\right) - \frac{k}{2}\left(k + \frac{55}{4}\right) + \frac{11}{2}\left(\frac{-5k}{4} - 11\right) = 0$$
$$\therefore 2k^2 + 55k + 350 = 0$$
$$\therefore (k+10)(2k+35) = 0$$
$$\therefore k = -10, \text{ or } -17\frac{1}{2}. \quad \blacksquare$$

§ **Problem 6.2.9.** $6x^2 + xy + ky^2 - 11x + 43y - 35 = 0$. ◊

§§ **Solution.** The given equation is
$$6x^2 + xy + ky^2 - 11x + 43y - 35 = 0.$$
By *Art.* 118, the Discriminant of this equation should be zero, i.e.,
$$\Delta = \begin{vmatrix} a & h & g \\ h & b & f \\ g & f & c \end{vmatrix} = 0.$$
where $a = 6$, $b = k$, $c = -35$, $h = \dfrac{1}{2}$, $g = -\dfrac{11}{2}$, $f = \dfrac{43}{2}$.

$$\therefore \Delta = \begin{vmatrix} 6 & \frac{1}{2} & -\frac{11}{2} \\ \frac{1}{2} & k & \frac{43}{2} \\ -\frac{11}{2} & \frac{43}{2} & -35 \end{vmatrix} = 0.$$

$$\therefore 6\left(-35k - \frac{1849}{4}\right) - \frac{1}{2}\left(-\frac{35}{2} + \frac{473}{4}\right) - \frac{11}{2}\left(\frac{43}{4} + \frac{11k}{2}\right) = 0$$
$$\therefore \frac{961}{4}k = -2883$$
$$\therefore k = -12. \quad \blacksquare$$

§ **Problem 6.2.10.** $kxy - 8x + 9y - 12 = 0$. ◊

§§ **Solution.** The given equation is
$$kxy - 8x + 9y - 12 = 0.$$
By *Art.* 118, the Discriminant of this equation should be zero, i.e.,
$$\Delta = \begin{vmatrix} a & h & g \\ h & b & f \\ g & f & c \end{vmatrix} = 0.$$
where $a = 0$, $b = 0$, $c = -12$, $h = \dfrac{k}{2}$, $g = -4$, $f = \dfrac{9}{2}$.

$$\therefore \Delta = \begin{vmatrix} 0 & \frac{k}{2} & -4 \\ \frac{k}{2} & 0 & \frac{9}{2} \\ -4 & \frac{9}{2} & -12 \end{vmatrix} = 0.$$

6.2. General Equation of The Second Degree 158

$$\therefore -\frac{k}{2}(-6k+18) - 4\frac{9k}{4} = 0$$
$$\therefore 3k^2 - 18k = 0$$
$$\therefore 3k(k-6) = 0$$
$$\therefore k = 0, \text{ or } k = 6.$$

But $k = 0$ will render the equation to first degree only, i.e. to a single straight line only, hence $k = 0$ is not correct here.
Hence $k = 6$. ∎

§ Problem 6.2.11. $x^2 + \dfrac{10}{3}xy + y^2 - 5x - 7y + k = 0$. ◊

§§ Solution. The given equation is
$$x^2 + \frac{10}{3}xy + y^2 - 5x - 7y + k = 0.$$
By *Art.* 118, the Discriminant of this equation should be zero, i.e.,
$$\Delta = \begin{vmatrix} a & h & g \\ h & b & f \\ g & f & c \end{vmatrix} = 0.$$
where $a = 1$, $b = 1$, $c = k$, $h = \dfrac{5}{3}$, $g = -\dfrac{5}{2}$, $f = -\dfrac{7}{2}$.
$$\therefore \Delta = \begin{vmatrix} 1 & \frac{5}{3} & -\frac{5}{2} \\ \frac{5}{3} & 1 & -\frac{7}{2} \\ -\frac{5}{2} & -\frac{7}{2} & k \end{vmatrix} = 0.$$
$$\therefore k - \frac{49}{4} - \frac{5}{3}\left(\frac{5k}{3} - \frac{35}{4}\right) - \frac{5}{2}\left(-\frac{35}{6} + \frac{5}{2}\right) = 0$$
$$\therefore k = 6. \quad \blacksquare$$

§ Problem 6.2.12. $12x^2 + xy - 6y^2 - 29x + 8y + k = 0$. ◊

§§ Solution. The given equation is
$$12x^2 + xy - 6y^2 - 29x + 8y + k = 0.$$
By *Art.* 118, the Discriminant of this equation should be zero, i.e.,
$$\Delta = \begin{vmatrix} a & h & g \\ h & b & f \\ g & f & c \end{vmatrix} = 0.$$
where $a = 12$, $b = -6$, $c = k$, $h = \dfrac{1}{2}$, $g = -\dfrac{29}{2}$, $f = 4$.
$$\therefore \Delta = \begin{vmatrix} 12 & \frac{1}{2} & -\frac{29}{2} \\ \frac{1}{2} & -6 & 4 \\ -\frac{29}{2} & 4 & k \end{vmatrix} = 0.$$
$$\therefore 12(-6k - 14) - \frac{1}{2}\left(\frac{k}{2} + 58\right) - \frac{29}{2}(2 - 87) = 0$$
$$\therefore \frac{289}{4}k = \frac{2023}{2}$$
$$\therefore k = 14. \quad \blacksquare$$

§ Problem 6.2.13. $2x^2 + xy - y^2 + kx + 6y - 9 = 0$. ◊

§§ Solution. The given equation is
$$2x^2 + xy - y^2 + kx + 6y - 9 = 0.$$

6.2. General Equation of The Second Degree

By *Art.* 118, the Discriminant of this equation should be zero, i.e.,
$$\Delta = \begin{vmatrix} a & h & g \\ h & b & f \\ g & f & c \end{vmatrix} = 0.$$
where $a = 2$, $b = -1$, $c = -9$, $h = \dfrac{1}{2}$, $g = \dfrac{k}{2}$, $f = 3$.
$$\therefore \Delta = \begin{vmatrix} 2 & \frac{1}{2} & \frac{k}{2} \\ \frac{1}{2} & -1 & 3 \\ \frac{k}{2} & 3 & -9 \end{vmatrix} = 0.$$
$$\therefore 2(9-9) - \frac{1}{2}\left(-\frac{9}{2} - \frac{3k}{2}\right) + \frac{k}{2}\left(\frac{3}{2} + \frac{k}{2}\right) = 0$$
$$\therefore \frac{9}{4} + \frac{3k}{4} + \frac{3k}{4} + \frac{k^2}{4} = 0$$
$$\therefore k^2 + 6k + 9 = 0$$
$$\therefore (k+3)^2 = 0$$
$$\therefore k = -3.$$
∎

§ Problem 6.2.14. $x^2 + kxy + y^2 - 5x - 7y + 6 = 0$. ◊

§§ Solution. The given equation is
$$x^2 + kxy + y^2 - 5x - 7y + 6 = 0.$$
By *Art.* 118, the Discriminant of this equation should be zero, i.e.,
$$\Delta = \begin{vmatrix} a & h & g \\ h & b & f \\ g & f & c \end{vmatrix} = 0.$$
where $a = 1$, $b = 1$, $c = 6$, $h = \dfrac{k}{2}$, $g = -\dfrac{5}{2}$, $f = -\dfrac{7}{2}$.
$$\therefore \Delta = \begin{vmatrix} 1 & \frac{k}{2} & -\frac{5}{2} \\ \frac{k}{2} & 1 & -\frac{7}{2} \\ -\frac{5}{2} & -\frac{7}{2} & 6 \end{vmatrix} = 0.$$
$$\therefore \left(6 - \frac{49}{4}\right) - \frac{k}{2}\left(3k - \frac{35}{4}\right) - \frac{5}{2}\left(-\frac{7k}{4} + \frac{5}{2}\right) = 0$$
$$\therefore 6k^2 - 35k + 50 = 0$$
$$\therefore (2k - 5)(3k - 10) = 0$$
$$\therefore k = \frac{5}{2}, \text{ or } \frac{10}{3}.$$
∎

§ Problem 6.2.15. *Prove that the equations to the straight lines passing through the origin which make an angle α with the straight line $y + x = 0$ are given by the equation*
$$x^2 + 2xy \sec 2\alpha + y^2 = 0.$$ ◊

§§ Solution. The slope of the line $y + x = 0$ is -1. Hence it is inclined at $\tan^{-1}(-1) = 135°$ to the x-axis.

Hence the slopes of the straight lines making an angle α with this line will be $\tan(135° + \alpha)$ and $\tan(135° - \alpha)$ respectively.

Since these lines passes through the origin, hence the equations to these lines will be
$$y = \tan(135° + \alpha)x, \text{ and}$$
$$y = \tan(135° - \alpha)x$$

6.2. General Equation of The Second Degree

Let us simplify these equations.
$$y = \tan(135° + \alpha)x = \frac{-1 + \tan \alpha}{1 + \tan \alpha}x$$
$$\therefore (\cos \alpha + \sin \alpha)y + (\cos \alpha - \sin \alpha)x = 0. \tag{6.8}$$

$$y = \tan(135° - \alpha)x = \frac{-1 - \tan \alpha}{1 - \tan \alpha}x$$
$$\therefore (\cos \alpha - \sin \alpha)y + (\cos \alpha + \sin \alpha)x = 0. \tag{6.9}$$

Combining the equations (6.8) and (6.9), we get the equations of the two straight lines as follows
$$\{(\cos \alpha + \sin \alpha)y + (\cos \alpha - \sin \alpha)x\}$$
$$\{(\cos \alpha - \sin \alpha)y + (\cos \alpha + \sin \alpha)x\} = 0$$
$$\therefore (x^2 + y^2)(\cos^2 \alpha - \sin^2 \alpha) + xy\left\{(\cos \alpha + \sin \alpha)^2 + (\cos \alpha - \sin \alpha)^2\right\} = 0$$
$$\therefore (x^2 + y^2)\cos 2\alpha + 2xy = 0$$
$$\therefore x^2 + 2xy \sec 2\alpha + y^2 = 0.$$

Alternative Solution :

The slope of the line $y + x = 0$ is -1.

Equation of the line passing through the origin is $y = mx$. If this is inclined at an angle α to the line $y + x = 0$, then
$$\tan \alpha = \frac{-1 - m}{1 - m}$$
$$\therefore m = \frac{1 + \tan \alpha}{\tan \alpha - 1} = \frac{\cos \alpha + \sin \alpha}{\sin \alpha - \cos \alpha}$$

With this value of m, the equation $y = mx$ translates to
$$(\cos \alpha - \sin \alpha)y + (\cos \alpha + \sin \alpha)x = 0. \tag{6.10}$$

If this is inclined at an angle $-\alpha$ to the line $y + x = 0$, then
$$-\tan \alpha = \frac{-1 - m}{1 - m}$$
$$\therefore m = \frac{\tan \alpha - 1}{1 + \tan \alpha} = \frac{\sin \alpha - \cos \alpha}{\cos \alpha + \sin \alpha}$$

With this value of m, the equation $y = mx$ translates to
$$(\cos \alpha + \sin \alpha)y + (\cos \alpha - \sin \alpha)x = 0. \tag{6.11}$$

Combining the equations (6.10) and (6.11), we get the equations of the two straight lines as already described in the previous solution. ∎

§ Problem 6.2.16. *What relations must hold between the coefficients of the equations*

(1) $ax^2 + by^2 + cx + cy = 0$, *and*

(2) $ay^2 + bxy + dy + ex = 0$,

so that each of them may represent a pair of straight lines ? ◊

§§ Solution. (i) The given equation is
$$ax^2 + by^2 + cx + cy = 0.$$

By *Art.* 118, the Discriminant of this equation should be zero, i.e.,
$$\Delta = \begin{vmatrix} a & h & g \\ h & b & f \\ g & f & c \end{vmatrix} = 0.$$

6.2. General Equation of The Second Degree

where $a = a$, $b = b$, $c = 0$, $h = 0$, $g = \dfrac{c}{2}$, $f = \dfrac{c}{2}$.

$$\therefore \Delta = \begin{vmatrix} a & 0 & \frac{c}{2} \\ 0 & b & \frac{c}{2} \\ \frac{c}{2} & \frac{c}{2} & 0 \end{vmatrix} = 0.$$

$$\therefore -a\dfrac{c^2}{4} + \dfrac{c}{2}\left(-\dfrac{bc}{2}\right) = 0$$

$$\therefore c(a+b) = 0.$$

(ii) The given equation is
$$ay^2 + bxy + dy + ex = 0.$$
By *Art.* 118, the Discriminant of this equation should be zero, i.e.,

$$\Delta = \begin{vmatrix} a & h & g \\ h & b & f \\ g & f & c \end{vmatrix} = 0.$$

where $a = 0$, $b = a$, $c = 0$, $h = \dfrac{b}{2}$, $g = \dfrac{e}{2}$, $f = \dfrac{d}{2}$.

$$\therefore \Delta = \begin{vmatrix} 0 & \frac{b}{2} & \frac{e}{2} \\ \frac{b}{2} & a & \frac{d}{2} \\ \frac{e}{2} & \frac{d}{2} & 0 \end{vmatrix} = 0.$$

$$\therefore \left(-\dfrac{b}{2}\right)\left(-\dfrac{de}{4}\right) + \dfrac{e}{2}\left(\dfrac{bd}{4} - \dfrac{ae}{2}\right) = 0$$

$$\therefore e(bd - ae) = 0$$

$$\therefore e = 0, \text{ or } ae = bd. \quad \blacksquare$$

§ Problem 6.2.17. *The equations to a pair of opposite sides of a parallelogram are*
$$x^2 - 7x = 6 = 0 \text{ and } y^2 - 14y + 40 = 0;$$
find the equations to its diagonals. ◊

§§ Solution. The given equations to first set of opposite sides are
$$x^2 - 7x + 6 = 0$$
$$\therefore (x-1)(x-6) = 0$$

Hence the equations are
$$x - 1 = 0, \text{ and}$$
$$x - 6 = 0.$$
Similarly, the given equations to second set of opposite sides are
$$y^2 - 14y + 40 = 0$$
$$\therefore (y-4)(y-10) = 0$$

Hence the equations are
$$y - 4 = 0, \text{ and}$$
$$y - 10 = 0.$$
Hence the coordinates of the vertices of the parallelogram are
$$(1,4); \ (6,4); \ (6,10); \ (1,10).$$
Hence the equation of the first diagonal is
$$y - 4 = \dfrac{10-4}{6-1}(x-1)$$
$$\therefore 5y - 6x = 14.$$

Similarly the equation of the second diagonal is
$$y - 4 = \frac{10-4}{1-6}(x-6)$$
$$\therefore 5y + 6x = 56.$$ ■

6.3 Equations Representing Isolated Points

§ Problem 6.3.1. *Prove that the equation*
$$y^3 - x^3 + 3xy(y-x) = 0$$
represents three straight lines equally inclined to one another. ◊

§§ Solution. The given equation is
$$y^3 - x^3 + 3xy(y-x) = 0.$$
By *Art.* 35, Transforming to polar coordinates by using $x = r\cos\theta$, $y = r\sin\theta$, the equation gets transformed to
$$\sin^3\theta - \cos^3\theta + 3\cos\theta\sin\theta(\sin\theta - \cos\theta) = 0$$

Dividing both sides by $\cos^3\theta$:
$$\tan^3\theta - 1 + 3\tan^2\theta - 3\tan\theta = 0$$
$$\therefore -1 = \frac{3\tan\theta - \tan^3\theta}{1 - 3\tan^2\theta} = \tan 3\theta$$
$$\therefore \tan 3\theta = -1 = \tan\left(-\frac{\pi}{4}\right), \text{ or } \tan\left(\pi - \frac{\pi}{4}\right), \text{ or } \tan\left(2\pi - \frac{\pi}{4}\right)$$
$$\therefore \theta = -\frac{\pi}{12}, \text{ or } \frac{3\pi}{12}, \text{ or } \frac{7\pi}{12},$$

The locus is therefore three straight lines passing through the origin and inclined at angles
$$-\frac{\pi}{12}, \frac{3\pi}{12}, \frac{7\pi}{12},$$
to the axis of x.

They are therefore equally inclined at angle $\frac{4\pi}{12} = \frac{\pi}{3}$ to one another.

Alternatively, we can get this solution by putting $m = -1$ in *Ex.* 2 of *Art.* 126.

Alternative Solution :

The given equation is
$$y^3 - x^3 + 3xy(y-x) = 0$$
$$\therefore (y-x)(y^2 + xy + x^2 + 3xy) = 0$$
$$\therefore (y-x)\left\{y + (2-\sqrt{3})x\right\}\left\{y + (2+\sqrt{3})x\right\} = 0.$$
This represent three straight lines with the equations as
$$y - x = 0 \tag{6.12}$$
$$y + (2-\sqrt{3})x = 0 \tag{6.13}$$
$$y + (2+\sqrt{3})x = 0. \tag{6.14}$$
Slopes of these lines are
$$m_1 = 1$$
$$m_2 = -(2-\sqrt{3})$$
$$m_3 = -(2+\sqrt{3})x.$$

6.3. Equations Representing Isolated Points

The angle between the lines (6.12) and (6.13) is given by

$$= \tan^{-1} \frac{m_1 - m_2}{1 + m_1 m_2}$$
$$= \tan^{-1} \frac{1 + 2 - \sqrt{3}}{1 - (2 - \sqrt{3})} = \tan^{-1} \frac{3 - \sqrt{3}}{\sqrt{3} - 1} = \tan^{-1} \sqrt{3} = \frac{\pi}{3}.$$

The angle between the lines (6.13) and (6.14) is given by

$$= \tan^{-1} \frac{m_2 - m_3}{1 + m_2 m_3}$$
$$= \tan^{-1} \frac{-(2 - \sqrt{3}) + (2 + \sqrt{3})}{1 + (2 - \sqrt{3})(2 + \sqrt{3})} = \tan^{-1} \frac{2\sqrt{3}}{2} = \tan^{-1} \sqrt{3} = \frac{\pi}{3}.$$

The angle between the lines (6.14) and (6.12) is given by

$$= \tan^{-1} \frac{m_3 - m_1}{1 + m_3 m_1}$$
$$= \tan^{-1} \frac{-(2 + \sqrt{3}) - 1}{1 - (2 + \sqrt{3})} = \tan^{-1} \frac{-(3 + \sqrt{3})}{-(\sqrt{3} + 1)} = \tan^{-1} \sqrt{3} = \frac{\pi}{3}.$$

They are therefore equally inclined at an angle $\frac{\pi}{3}$ to one another. ∎

ॐ

This page is intentionally left blank.

www.ingramcontent.com/pod-product-compliance
Lightning Source LLC
Chambersburg PA
CBHW020658220526
45464CB00001B/493